**DIGGING UP THE BONES®**

MEDICAL
REVIEW
SERIES

# *Anatomy*

# Anatomy

Compiled and Written by
## Nikos M. Linardakis, M.D.
*Editor-in-Chief*
*Digging Up the Bones Series*

## McGraw-Hill
*Health Professions Division*

New York  St. Louis  San Francisco  Auckland  Bogotá  Caracas  Lisbon  London
Madrid  Mexico City  Milan  Montreal  New Delhi  San Juan  Singapore  Sydney
Tokyo  Toronto

# McGraw-Hill

*A Division of The McGraw-Hill Companies*

**ANATOMY: Digging Up the Bones**

1234567890 MALMAL 99

ISBN 0-07-038415-0

This book was set in Times Roman by V & M Graphics, Inc.
The editors were John Dolan and Steven Melvin;
the production supervisor was Helene G. Landers;
the cover designer was Mathew Dvorozniak;
illustrations and graphic assistance by
 Eric Melander and Nikos M. Linardakis, M.D.
Malloy Lithographing, Inc., was the printer and binder.

This book is printed on acid-free paper.

**Library of Congress Cataloging-in-Publication Data**

Linardakis, Nikos M.
    Anatomy / compiled and written by Nikos M. Linardakis;
 illustrations and graphic assistance by Eric Melander and
 Nikos M. Linardakis.
        p.    c.m. — (Digging up the bones)
     ISBN 0-07-038415-0
     1. Human anatomy.    2. Anatomy, Pathological.    I. Title.
 II. Series: Digging up the bones medical review series.
     [DNLM: 1. Anatomy.  QS 4  L735a  1999]
 QM23.2.L54     1999
 611—dc21
 DNLM/DLC
 for Library of Congress                                    99-24495
                                                                CIP

Το Παππου Νικο,

Enjoy, from your "grandson,"

Ιατρος Νικο Λιναρδακης

# Contents

# Preface

This volume of *Digging up the Bones® Medical Review Series* is a compilation summary of over a thousand exam questions covering the past 10 years which were frequently seen in Anatomy. In the *Digging Up the Bones®* review series, we will cover several review items that are essential for USMLE I and course exam review. I have italicized and boldfaced the key facts that have been heavily tested in the past. I believe the following will be an enormous tool in "digging up the bones" of anatomy, and to increase your understanding (and of course your grade!). I recommend placing your own notes on the side margins, and review this material at least twice to understand and MEMORIZE it. Try to create clinical scenarios to help in this memorization process. Overall, the information presented should be of tremendous help.

This volume of the *Digging Up the Bones®* medical review series was created to present Anatomy highlights in a straight-forward, easy-to-learn format. Please take the time to learn ALL of the illustrations and noted anatomical landmarks and organs. It is important to recognize the neighboring organs and landmarks as they are positioned in the body. On your examinations, you will be asked questions in reference to their location in the body, and will be tested on your understanding of the structure's interdependence on other structures.

This book is presented in regions from the head to the lower extremity. It is an informative review with several clinical correlations.

*Digging Up the Bones® Medical Review Series* has received recognition from around the world through the shared comments with physicians and students. Your mastery of these topics should improve your clinical skills. I know your efforts will pay off, and you will share these skills with others in the years to come.

Now, let's get started!


Nikos M. Linardakis, M.D.

# Introduction to Gross Anatomy

1

## SKELETAL, MUSCULAR, VASCULAR, AND LYMPHATIC SYSTEMS

The bones of an individual are made of organic matter. Bones develop the most in children. With a deficiency of calcium (and lack of calcification), rickets will develop in a child (on x-ray look for enlarged ends of the ribs 2 through 8 creating rachitis—rosary bead–like enlargements) and in an adult, osteomalacia will develop. Children have the greatest number of fractures. This is due to their thin bones and increased activity.

With advanced age, osteoporosis may occur as a result of a decrease in the bone matter. This occurs most commonly with women. Because the nerves travel with the arteries within the bones, if there is a fracture, there is great pain. The periosteum is also very sensitive to any tension. The other degenerative changes that can occur are bone and joint inflammation within synovial joints. Degenerative joint disease is also considered osteoarthritis (presents with stiffness and pain) and usually affects the joints of elderly people. A surgical procedure to visualize the problems of hip or knee joints is called arthroscopy.

Skeletal muscle develops the movement necessary for the skeleton. Increased muscle tone is considered *spasticity*, and decreased muscle tone is considered *flaccidity*. Decreased muscle tone occurs with anesthesia. If there is a nerve that is severed, or denervation, in a muscle group, look for muscle atrophy (decrease or loss of muscle mass). Following muscle damage, scar tissue may replace damaged muscle. In the myocardium, or heart muscle, damaged muscle which turns to scar tissue will become necrotic. This is known as a myocardial infarction (MI). Hypertrophy is the growth of muscle in the size (NOT numbers) of the muscle fibers. An example is hyperplasia (an increase in number of smooth muscle cells), and it is seen in changes during pregnancy.

Atherosclerosis is the fibrosis and calcification that occurs in the vascular system and may result in ischemic heart disease. Myocardial infarction, strokes, and gangrene are other complications of these vascular changes. Thrombosis is a clotting of the blood that forms an occlusion of the arteries as atherosclerosis. Once the thrombotic lesion dislodges from the vessel wall, it is considered an embolus. Varicose veins are due to dilated veins caused by loss of the elasticity as well as incompetent valves. Arteries, veins, and capillaries are summarized in the following table.

| Arteries | Veins | Capillaries |
|----------|-------|-------------|
| These vessels carry the blood *away* from the heart *to* the body. There are three layers: the tunica *intima*, tunica *media*, and tunica *adventitia*. The three types of arteries are: *elastic* (the largest vessels; i.e., aorta), *muscular* (distributing arteries, circular smooth muscle), and *arterioles* (smallest arteries, maintain the blood pressure). | These vessels *return* the blood back to the heart from the capillaries. There are three types: *small, medium* (contain valves which allow the blood to go forward, but close the reverse flow off), and *large* (SVC, or superior vena cava, is an example). | These vessels are *simple endothelial* vessels that join the arterial and venous systems. The network of capillaries (*capillary bed*) joins the arterioles to the venules and allows for the exchange of nutrients and oxygen with the surrounding tissues. |

## LYMPHATIC SYSTEM

The lymphatics are vessels that connect with lymph nodes and the spleen (a lymphatic organ). The tissue fluids filter through the lymphatics. The lymph flows through the lymph nodes and then through lymphatic trunks to end up in the thoracic duct. This then climbs into the internal jugular and subclavian veins or through the lymphatic ducts. The superficial lymphatic vessels of the skin drain into the deep lymphatic vessels. The lymphatics drain the tissue fluid (plasma) from the venous system, absorb and transport fat, and create an immunologic, lymphocytic system that is important for the wellbeing of an individual. Lymphangitis is inflammation of the lymphatic vessels. Lymphadenitis is inflammation of the lymph nodes. In cancer, the cancerous spreading can cause collection and blockage in lymph vessels, which may lead to lymphedema, or the accumulation of fluid (Fig. 1-1).

In the growth of a long bone, there is a formation of primary and secondary centers of ossification. That is, the long bones grow through endochondral ossification. The growth in the length of the bone occurs from each side of the epiphyseal plates. The diaphysis makes up the body, or shaft, of the bone— and is calcified from a primary ossification center. The epiphyses are located at the ends of the bone (with the epiphyseal plate between the diaphysis and

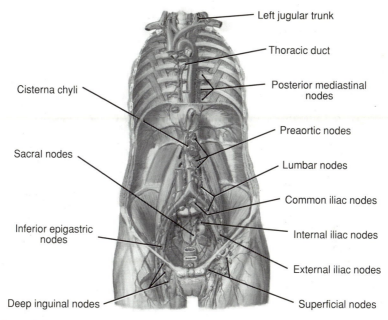

Figure 1-1 Lymph vessels and nodes of posterior abdomen.

epiphysis) and are formed after calcification of secondary ossification centers (secondary because they usually appear after birth). The types of muscles and their derivation are listed in the tables that follow.

| Skeletal Muscle | Cardiac Muscle | Smooth Muscle |
|---|---|---|
| This muscle allows the movement of the *skeleton*. It is usually *voluntary* (you can control the movement), but some are involuntary or automatic (i.e., the diaphragm is automatic until you want to get some forced air in). It has at least two attachments (to bone, skin, fascia, or other sites). | This muscle forms the majority of the *myocardium*, or heart muscular wall. It is composed of *striated* cardiac muscle that is NOT voluntary (instead, a *pacemaker* innervates the fibers through the *autonomic* nervous system). | This muscle is found inside the *tunica media*, or middle wall of the blood vessels, and the digestive tract (*peristalsis* is the wave-like movement created through rhythmic smooth muscle contractions to forward items through the tube). This is also an *autonomic* nervous system controlled *involuntary* muscle. |

| Derived from Embryonic | Forms |
|---|---|
| Notochord | Cells of the nucleus pulposus (inner portion of the intervertebral disk) |
| Blood vessels near the heart | Atrial internal smooth-walled area |
| Primitive heart | Pectinate muscles, atrial internal rough-walled area |
| | Trabeculae carneae, ventricular rough-walled area |

# Notes

# Nervous Systems 2

---

## CENTRAL NERVOUS SYSTEM

The central nervous system, or CNS, is composed of the brain and the spinal cord. The CNS allows neurologic signals to be understood and implemented at an advanced level for coordination, thought, and other aspects of learning. The membranes that protect the brain with cerebrospinal fluid are the meninges (known as the dura mater, pia mater, and arachnoid mater).

Learn all of the areas of the brain in the illustrations provided in Figs. 2-1 and 2-2.

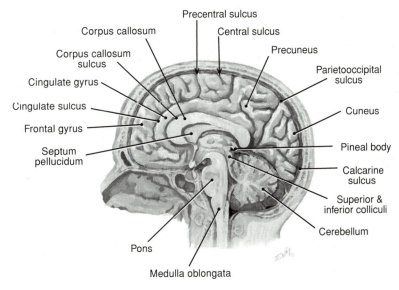

**Figure 2-1**   Cerebrum: sagittal section of the brain.

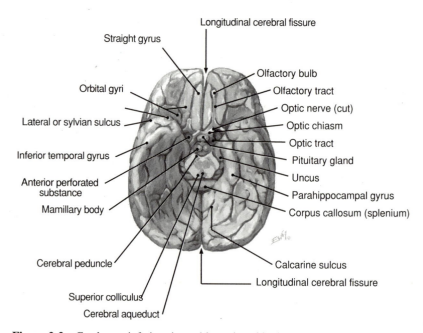

**Figure 2-2** Cerebrum: inferior view with sectioned brain stem.

## PERIPHERAL NERVOUS SYSTEM

The peripheral nervous system, or the PNS, is composed of the cranial nerves (12 nerve pairs that originate from the *brain*) and the spinal nerves (31 nerve pairs that originate from the *spine*). Sensory organs in the *periphery* bring impulses to the CNS that then travel back to the appropriate muscles (or glands) for their actions. Each nerve *axon* is covered with a myelin sheath and is then covered with *endoneurium* (around the nerve fiber). A bundle of nerve fibers (*fasciculus*) is bundled together with *perineurium*, and *epineurium* covers the entire nerve.

## SOMATIC NERVOUS SYSTEM

The somatic nervous system contains a *sensory* (sense of touch, pain and temperature, and position) and a *motor* division (for voluntary muscle contractions). There is a *somatic* portion to the CNS and a *somatic* portion to the PNS.

## AUTONOMIC NERVOUS SYSTEM

The body uses an integrated system to operate, which includes the *autonomic* and *somatic* nervous systems. The **somatic nervous system** innervates the *skeletal muscles* with the use of a *single* neuron. The **autonomic nervous** *system* innervates the *smooth muscles, cardiac muscles,* and the *glands* with the use of *two neurons*. These are the *preganglionic* neuron (with a cell body within the *central* nervous system) and the *postganglionic* neuron (with a cell body in the *peripheral* nervous system). Remember, a neuron is composed of a cell body and its processes (axon and dendrites).

The autonomic nervous system offers innervation through two neurons, with the exception of the adrenal medulla (one neuron). The autonomic nervous system is made of two systems, the sympathetic and the parasympathetic nervous systems (Fig. 2-3). In general, the sympathetic nervous system

CNS:                    PNS:

Sympathetic system:

1° neuron (preganglionic)

(short)

2° postganglionic

neuron

(long)

end organ

Parasympathetic system:

1° neuron (preganglionic)

(long)

2° postganglionic

neuron

(short)

end organ

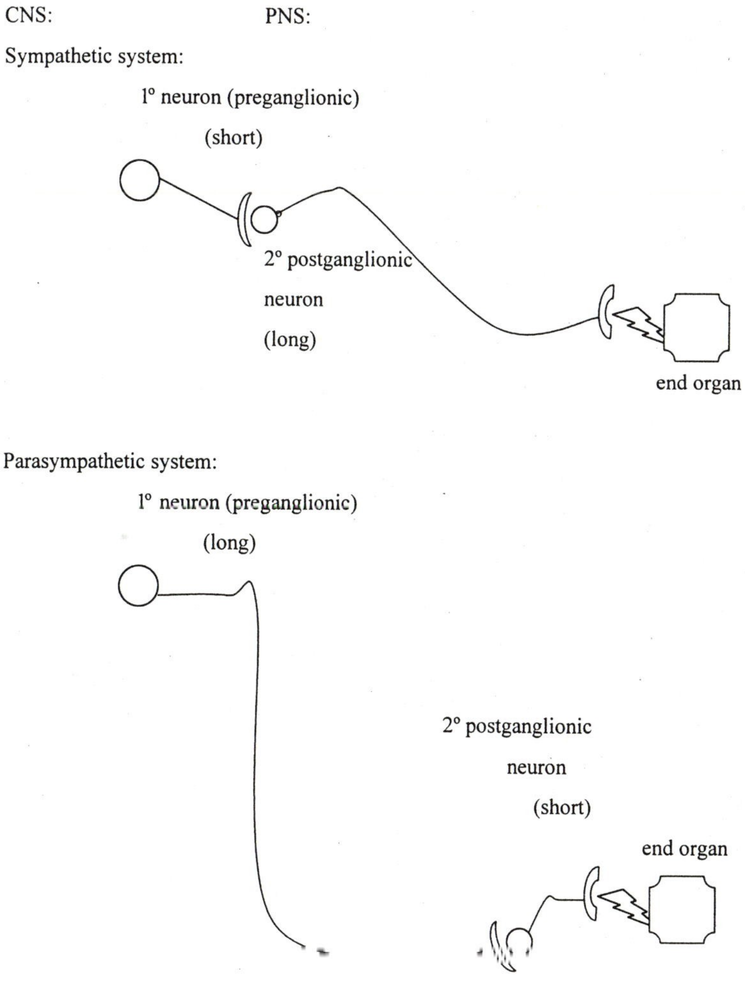

**Figure 2-3**

preganglionic neuron cell bodies are in the thoracolumbar region (T1–L2). The sympathetic postganglionic cell bodies are in the peripheral nervous system near the CNS, but away from the end organ. Therefore, the preganglionic axons are short and the postganglionic axons are long in the sympathetic system. The sympathetic system is known as the "fight or flight" system as it increases the heart rate and force of contraction, dilates the pupils, and decreases gastrointestinal motility. ONLY the sympathetic system innervates the sweat glands, the erector pilar muscles of hair, and cutaneous vascular smooth muscle. The parasympathetic nervous system preganglionic neuron cell bodies are in the CNS at the cranial-sacral regions (S2–S4 and cranial nerves III, VII, IX, and X). The postganglionic cell bodies are in the PNS and far away from the CNS, but close to the end organ. Therefore, the preganglionic axons are long and the postganglionic axons are short in the parasympathetic system. The parasympathetic system decreases the heart rate, increases gastrointestinal motility, constricts the pupils and airway, and decreases sweat production.

---

### Horner's Syndrome

Denervation of the sympathetic system in the cranial area (traveling along the carotid arteries) leading to (1) ptosis (drooping eyelid), (2) miosis (pupillary constriction), and (3) anhidrosis (inability to sweat).

# Notes

# The Head and Neck

3

## THE HEAD

The head has a few important areas that must be learned. These include the zygomatic arch, canals, and sutures. The zygomatic arches are the widest location of the face and are often the location of a fracture. Subsequently, because of the location (in proximity to the eye), a fractured zygomatic arch may cause injury to the eye. In addition, this may fracture the nasal bones. Upon impact, the skull may be fractured at the *opposite end*. This is known as a *countercoup fracture*. The olfactory nerve passes through the *cribriform plate* (in the anterior cranial fossa). A few other fossae that carry important cranial nerves include:

1. The *foramen rotundum*, which carries the *maxillary* division of the trigeminal nerve ($V_2$).
2. The *foramen ovale*, which carries the *mandibular* division of the trigeminal nerve ($V_3$). In addition, there is the lesser petrosal nerve and the accessory meningeal artery.
3. The *jugular foramen*, which passes cranial nerves IX, X, and XI.
4. The *foramen magnum*, which passes the spinal accessory nerve (XI).

## VASCULATURE OF THE BRAIN

The brain (Fig. 3-1) is supplied by four main arteries, the two *vertebral arteries* (which supply mainly the brain stem, cerebellum, and occipital lobe) and the two *internal carotid arteries* (which take care of the rest of the cerebrum). Anastomosis occurs at the *circle of Willis*. The vertebral arteries are joined and form the *basilar artery*. The basilar artery then divides into the *posterior cerebral arteries*. The internal carotid artery (ICA) makes a division to form the *anterior cerebral artery*, the *middle cerebral artery*, and the *posterior communicating artery* (which connects the carotid and vertebral systems; ICA to the *posterior cerebral artery*). As such, the *anterior cerebral arteries* are connected by the *anterior communicating artery* (connects the left and right carotids). The arteries that supply the brain are shown in Fig. 3-2.

13

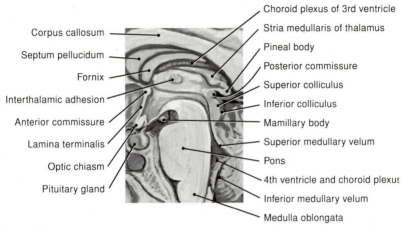

Corpus callosum

Septum pellucidum

Fornix

Interthalamic adhesion

Anterior commissure

Lamina terminalis

Optic chiasm

Pituitary gland

Choroid plexus of 3rd ventricle

Stria medullaris of thalamus

Pineal body

Posterior commissure

Superior colliculus

Inferior colliculus

Mamillary body

Superior medullary velum

Pons

4th ventricle and choroid plexus

Inferior medullary velum

Medulla oblongata

**Figure 3-1**   Brain: close-up view.

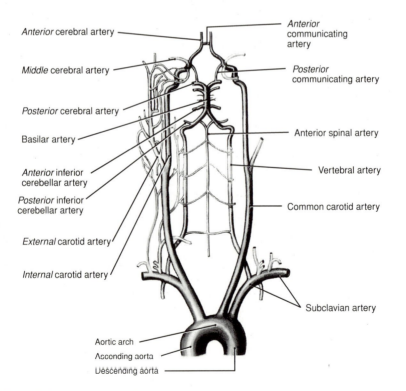

Anterior cerebral artery

Middle cerebral artery

Posterior cerebral artery

Basilar artery

Anterior inferior
cerebellar artery

Posterior inferior
cerebellar artery

External carotid artery

Internal carotid artery

Anterior
communicating
artery

Posterior
communicating artery

Anterior spinal artery

Vertebral artery

Common carotid artery

Subclavian artery

Aortic arch

Ascending aorta

Descending aorta

**Figure 3-2**   Arteries to the brain.

14

## THE SKULL

The skull forms a protection around the brain and the brain stem and organizes the organs for important senses (i.e., eyes, ears, tongue, and nasal structures). There are several skull landmarks that must be learned. Please refer to the illustrations in Figs. 3-3 through 3-6 for the newborn and the human skull.

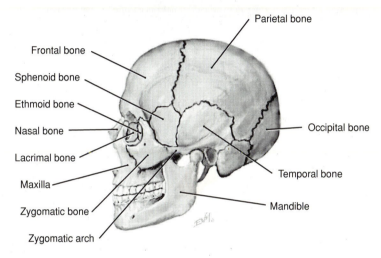

**Figure 3-3**    Lateral view of skull.

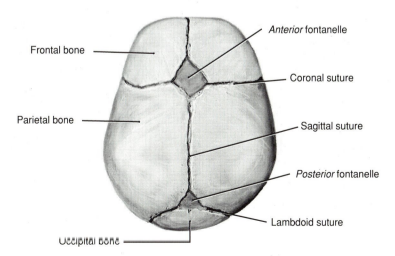

**Figure 3-4**    Newborn skull.

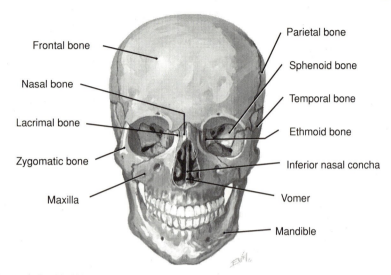

Frontal bone

Nasal bone

Lacrimal bone

Zygomatic bone

Maxilla

Parietal bone

Sphenoid bone

Temporal bone

Ethmoid bone

Inferior nasal concha

Vomer

Mandible

**Figure 3-5**  Skull bones.

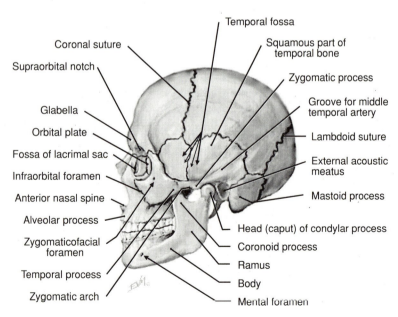

Temporal fossa

Coronal suture

Supraorbital notch

Squamous part of
temporal bone

Zygomatic process

Groove for middle
temporal artery

Glabella

Orbital plate

Fossa of lacrimal sac

Infraorbital foramen

Anterior nasal spine

Alveolar process

Zygomaticofacial
foramen

Temporal process

Zygomatic arch

Lambdoid suture

External acoustic
meatus

Mastoid process

Head (caput) of condylar process

Coronoid process

Ramus

Body

Mental foramen

**Figure 3-6**  Skull landmarks.

The *sagittal* suture is midline with a *coronal* suture, which angles at the *bregma* of the skull. The coronal suture separates the parietal bone from the frontal bone. Notice the other sutures and their location in relation to the bones and structures.

## DURA AND SINUSES

The *periosteal dura* lines the inner surface of the skull. *Meningeal dura* is fused to this periosteal dura until parts where it separates to bend inward, forming the *dural septa*. The major septa include the: *falx cerebri, falx cerebelli*, and *tentorium cerebelli*. The venous sinuses are between the *periosteal dura* and the *meningeal dura* or in parts of the meningeal dura. The sinuses and dural septa are illustrated in Fig. 3-7; become familiar with their relative location to other structures in the brain. For example, the *superior sagittal sinus* is located between the *periosteal* and *meningeal dura* at the base of the *falx cerebri*. The dural sinuses carry the majority of the venous drainage from the brain. This eventually drains to the *internal jugular vein* and *facial, scalp,* and *neck emissary veins*. These vessels do NOT contain valves, and therefore, the blood can enter in or out of the cranial cavity—making *emissary veins* a likely pathway for infection.

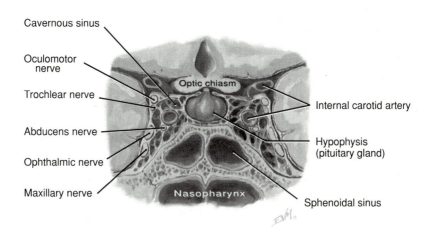

**Figure 3-7**   Cavernous sinus (coronal section).

## THE EYE

Let us review the orbit walls and the openings. The *roof* of the orbit is mainly the *frontal bone* (notice the frontal sinus and anterior cranial fossa). The *floor* is mainly the *maxilla*. The *medial* side is the *ethmoid* and *sphenoid bones* with the *nasal cavity*. The *lateral* wall is mainly the *zygoma* and the *sphenoid bones*. The orbit has a posterior opening, the *optic canal*, as well as the superior and inferior orbital fissures.

There are seven extraocular eye muscles to be aware of. The *levator palpebrae superioris* attaches to the **upper** eye*lid*, which elevates the upper eyelid (the other muscles attach to the eye*ball*). The *superior rectus* elevates and **add**ucts the eye, and the *inferior rectus* depresses and also **add**ucts the eye. Then, the *lateral rectus* **ab**ducts the eye, and the *medial rectus* **add**ucts the eye. The *superior oblique* is responsible for the depression and abduction of the eye, and the *inferior oblique* for the elevation and abduction of the eye. Innervation to the lateral rectus is by the *abducens nerve* (CN VI). The superior oblique muscle is innervated by the *trochlear nerve* (CN IV), and the remaining eye muscles receive innervation from the *oculomotor nerve* (CN III). The only exception is the levator palpebrae superioris, which is made of two parts; a *skeletal* muscle part is innervated by the oculomotor nerve (CN III) and a *smooth* muscle part is innervated by *sympathetic* nerves (Fig. 3-8).

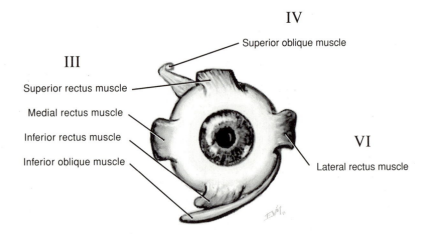

**Figure 3-8** Extrinsic eye muscles. III, oculomotor nerve; IV, trochlear nerve; VI, abducens nerve.

## COMMON INJURIES AND LESIONS OF THE HEAD

### CRANIAL NERVE III

In *third nerve palsy*, the upper eyelid droops and it will NOT be able to rise. This is the result of damage to the *superior* division of this cranial nerve. It affects the movement of the *levator palpebrae superioris.*

### TRIGEMINAL NERVE

Trigeminal *neuralgia* is the occurrence of sudden attacks of pain simply by touching the various areas of the trigeminal nerve divisions. A lesion of the trigeminal nerve results in the *anesthesia* of the *anterior half* of the scalp, the face (excluding the mandibular angle), and the various mucous membranes (nose, mouth, and anterior two thirds of the tongue). The trigeminal nerve may have some effect on the muscles of mastication (atrophy).

### FACIAL NERVE INJURY

Injury to the facial nerve (cranial nerve VII) can result in paresis or paralysis of the facial muscles on the *ipsilateral* (or affected) side—also known as *facial palsy*. You may already be aware of *Bell's palsy*. This is the paralysis of the facial nerve for no apparent reason (for example, a cold draft). The result is an inflammation of cranial nerve VII with edema and compression of the nerve fibers. The individual with Bell's palsy has difficulty closing his or her lips and eyelids (on the *ipsilateral* side). In addition, because the buccinator muscle is not able to fully perform, the individual will not be able to whistle (instead he or she will "withle") and will dribble. The facial nerve enters the *parotid gland* and carries with it the *parasympathetic* fibers. You will recall that in *mumps* the parotid gland may become infected, causing major pain and blocking off the parotid duct. This may create deposits that block the duct.

### HYPOGLOSSAL NERVE

Paralysis to the hypoglossal nerve can result in *ipsilateral* atrophy of the tongue. This causes the tongue to deviate to the *ipsilateral* side of the lesion when the tongue is protruded.

### HYDROCEPHALUS

The *obstruction* of cerebrospinal fluid (CSF) *flow*, interruption of CSF *absorption*, or *increased production* of CSF will increase intracranial pressure (ICP).

### EPIDURAL HEMORRHAGE

Bleeding that occurs between the *dura* and the *internal periosteum* is considered to be an *epidural hemorrhage*. It can result from a concussion or trauma to the head—among other causes.

### SUBDURAL HEMORRHAGE

Bleeding that occurs in the dura following brain movement from a blow to the head, or as seen with a ruptured *superior cerebral vein*, or in the elderly, should be considered in the diagnosis of *subdural hemorrhage*.

### SUBARACHNOID HEMORRHAGE

This is bleeding that occurs in the *subarachnoid space* following a ruptured aneurysm of the intracranial artery. This is often seen following a skull fracture or laceration.

### INTRACEREBRAL HEMORRHAGE

When there is bleeding *into* the brain (possibly from the *middle cerebral artery*), then we may consider *intracerebral hemorrhage*. It is commonly accompanied by hypertension or a paralysis. A *stroke* can result due to insufficient blood flow to the brain. This is known as a *cerebrovascular accident*, or CVA. It may be seen when a large artery (*internal carotid*) becomes occluded and there is an instant intracerebral hemorrhage.

### CSF RHINORRHEA

A CSF leak following a tear in the *meninges* as a result of a fractured *cribriform plate* is termed CSF rhinorrhea.

### MIDDLE MENINGEAL ARTERY

The *middle meningeal* artery is the most commonly injured artery in head trauma.

---

## THE TONGUE

The tongue is composed of intrinsic and extrinsic skeletal muscles (genioglossus, hyoglossus, and palatoglossus). The *hypoglossal nerve* (CN XII) innervates the tongue except for the palatoglossus, which is innervated by the *vagus nerve* (CN X). The sensory innervation to the mucosa of the tongue is provided by cranial nerves V, VII, IX, and X; with the anterior two thirds of the tongue receiving general sensory innervation from the *mandibular* division of the *trigeminal* nerve (CN $V_3$) and taste sensation from the *facial* nerve (CN VII). The *posterior* one third receives the general sensory *and* taste sensations from the *glossopharyngeal* nerve (CN IX). Toward the epiglottis area, the general sensory and taste function is from the *vagus* nerve (CN X).

---

## MUSCLES OF MASTICATION

The muscles of mastication are the *masseter, temporalis, lateral,* and *medial pterygoid* muscles. These four muscles receive innervation from the *mandibular branch of the trigeminal nerve* (CN $V_3$). The *lateral pterygoid* is the only muscle which depresses the mandible (or *opens* the mouth). The others are responsible for elevating the mandible (closing the mouth is more of an active process).

## THE EAR

The ear has three divisions, the *external, middle,* and *inner* ear. The external ear is the "collector" of sound waves and is composed of the *pinna* and the *external ear canal.* Notice that the outer canal is cartilage and the inner is bone; it directs the sound to the *tympanic membrane.* This membrane separates the external ear from the next area, the *middle ear.*

The middle ear is the "conductor" or amplifier of the sound waves. It has three *ossicles* called the *malleus* (which is attached to the inner part of the tympanic membrane), the *incus,* and the *stapes* (which has a portion within the *oval window* of the *inner ear*). Sound causes vibrations of the tympanic membrane and then the ossicles vibrate through the inner ear. The *tensor tympani* (attached to the **m**alleus) and the *stapedius* (attached to the **stape**s) are two muscles which decrease the amplitude of vibration of the membrane and ossicles during episodes of loud noise. This is a protective measure for the cochlear hair cells. A middle ear infection can result from infection of the pharynx to the middle ear (through the nasopharynx, by virtue of the eustachian auditory tube). *Hyperacusis* may result if the facial nerve (CN VII) is injured and the stapedius loses its function.

Finally, the inner ear (Fig. 3-9, *see page 22*) contains the auditory refinement equipment and balance. It is the "studio." The *cochlea* is responsible for auditory sensation, and the *semicircular canals, utricle,* and *saccule* are responsible for *vestibular* sensation. The cochlea contains a duct (filled with *endolymph*) and the *organ of Corti.* This organ has the special *cochlear hair cells. Perilymph* surrounds the duct and touches the stapes at the oval window. Vibration causes vibration of the organ of Corti. The hair cells convert this to a nerve impulse through the *cochlear nerve* (CN VIII). For vestibular function, head movement displaces the *vestibular hair cells* and creates a nerve impulse to the *vestibular nerve* (CN VIII).

## THE NECK

The neck presents several muscles that are unique to the development of facial expression (Fig. 3-10). The *platysma* muscles offer the muscles of "facial expression." The supply to the platysma is the *cervical branch* of the *facial nerve.* The *sternocleidomastoid* (SCM) muscle can be injured during birth of an infant with a breech presentation. This can result in *congenital torticollis,* with a recognizable tilt of the head as a result of a *contracted SCM*—the head is *pulled* to the ipsilateral side and the facial chin is upwardly rotated to the opposing side.

There are two types of fascia; superficial (which has loose connective tissue and fat) and deep fascia. Inside the superficial fascia is the previously mentioned *platysma* muscle, as well as superficial veins (external and anterior jugular veins) and cervical plexus cutaneous branches—which reach the back scalp and skin of the neck. The deep fascia is composed of several layers and contains the trapezius and sternocleidomastoid (SCM) muscles as well as the major enclosures for the common carotid and internal carotid arteries and the internal jugular vein and the vagus nerve.

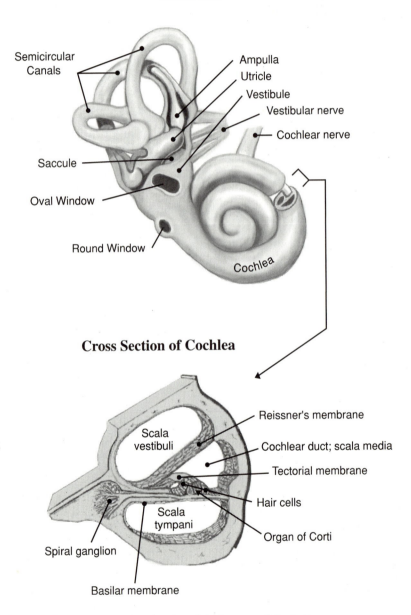

**Figure 3-9**  Schematic representation of the inner ear (labyrinth).

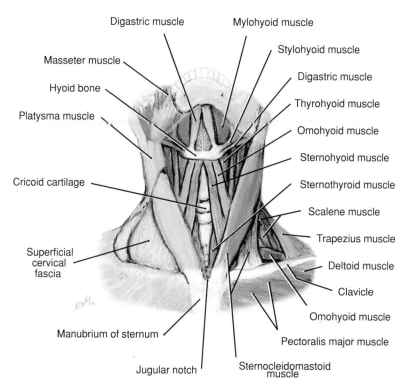

**Figure 3-10**    Neck muscles.

---

## COMMON INJURIES OR LESIONS

### ACCESSORY NERVE

Injury to the eleventh cranial nerve (CN XI), or *accessory nerve*, can create a problem for the individual in turning his or her head in opposition to the examiner's resistance (test the accessory nerve function by asking the patient to apply the force of his or her face against your hand). In the case of a paralysis to the *ipsilateral trapezius muscle*, the individual cannot lift or retract the shoulder (shoulder drooping) and has problems with lifting the arm above the shoulder level.

## PHRENIC NERVE
If the *phrenic nerve* is injured, it will result in the *paralysis* of *half* of the diaphragm.

## EXTERNAL LARYNGEAL NERVE
Injury to the *external laryngeal nerve* can result in a *monotonous* voice as a result of paralysis of the *cricothyroid muscle*.

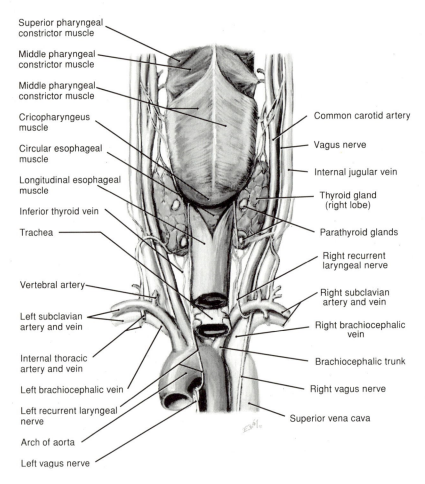

**Figure 3-11**   Posterior view of thyroid and pharynx

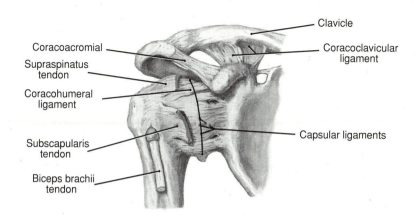

**Figure 3-12**  Glenohumeral joint: anterior view.

## RECURRENT LARYNGEAL NERVE

Injury to the *recurrent laryngeal nerve* can cause *dysphagia* (inability to eat and choking on food) as a result of paralysis of the muscles of the epiglottic folds. The individual most likely will NOT be able to speak.

## HORNER SYNDROME

Injury to the *superior cervical ganglion* at the ipsilateral *sympathetic pathway* results in Horner syndrome. An injury to the spinal cord, at the *cervical* area, may cause Horner syndrome. If the sympathetic trunk is lacerated in the neck, the following will result.

1. Pupillary constriction—from paralysis of the dilator pupillae muscle of the iris.
2. *Vasodilation* and *anhidrosis*—inability to sweat at the face and neck due to the lack of sympathetic supply.
4. Eyeball sunken into the orbit—because of the *orbitalis muscle*.

# Notes

# The Cranial Nerves    4

---

## CRANIAL NERVES (CN)

There are twelve (12) cranial nerves which are connected to the CNS. The cranial nerves are sensory, motor, or mixed (sensory and motor functions) (Fig. 4-1).

| **Purely Sensory** | **Purely Motor** | **Mixed (Sensory and Motor)** |
|---|---|---|
| Optic (II) | Oculomotor (III) | Trigeminal (V) |
| Olfactory (I) | Trochlear (IV) | Facial (VII) |
| | Abducens (VI) | Glossopharyngeal (IX) |
| | Accessory (XI) | Vagus (X) |
| | Hypoglossal (XII) | |

**Figure 4-1**    Cranial nerve components: sensory, motor, or mixed.

### CN I: THE OLFACTORY NERVE

The olfactory nerve is responsible for *smelling*, or *olfaction*. The *olfactory nerve* consists of **bi**polar neurons with filaments exiting from the cranial cavity through the *cribriform plate* of the *ethmoid bone*, which then synapse with the mitral cells at the *olfactory bulb*. Here, the olfactory bulb contains the neurons that join to form the olfactory *tract*. The olfactory tract passes through the *uncus* of the brain. If there is a lesion in the *uncus* of the *temporal lobe*, the individual may experience *olfactory* hallucinations.

An olfactory nerve lesion will result in *anosmia*, the loss of the sense of *smell*. Causes of such a lesion include: frontal lobe tumor (near the uncus) or fracture of the cranial fossa at the *cribriform plate* (check to see if the individual has a runny nose—this may be CSF rhinorrhea or an infection).

An optic nerve lesion will result in *total blindness*. It will also cause a loss of the *pupillary light and accommodation reflexes* on the *affected* side.

Increased intracranial pressure may impinge the optic nerve and create blindness.

*Papilledema* can occur when the optic nerve fibers enlarge (because the central retinal vein is engorged and the intracranial pressure is high). The optic disk, or *papilla*, swells from edema and increased ICP. This causes visual field defects.

If the *optic chiasma* is affected, consider a *pituitary tumor* that compresses the nerve fibers.

**Figure 4-2** Lesions and conditions of the optic pathway.

## CN II: THE OPTIC NERVE

The optic nerve is responsible for *sight*, or *vision*. The optic nerve contains the visual fibers of the retina (temporal and nasal half). This cranial nerve exits the orbit through the optic canal to form the *optic chiasma*—a crossing of the fibers occurs at this location (the nasal fibers cross while the fibers of the lateral half do NOT cross). This will relay the visual content and pupillary light reflex for vision. Figure 4-2 demonstrates conditions that will occur with a lesion along a particular area of the optic pathway.

## CN III: THE OCULOMOTOR NERVE

The oculomotor nerve is responsible for several eye movements and parasympathetics to the sphincter pupillae and ciliary muscles. The oculomotor nerve exits at the brain stem and follows through the *cavernous sinus* to advance into the *superior orbital fissure* and the orbit of the eye. Then, the nerve divides into the *superior* and *inferior oculomotor nerve rami*. The nerve possesses autonomic (somatic—superior ramus, and visceral—inferior ramus) components and innervates the *extraocular muscles* (but NOT the *superior oblique* or the *lateral rectus muscles*). This innervates the *constrictor pupillary* and ciliary muscles.

An oculomotor nerve lesion will result in *oculomotor palsy*:

*Ptosis*, or *drooping* of the upper eyelid, occurs secondary to the paralysis of the *levator palpebrae muscle*.

*Mydriasis* is the *dilation* of the pupil, and it occurs secondary to the paralysis of the *constrictor pupillary muscle*.

*Lateral strabismus*, or squinting, can throw off the *parallelism* of the visual axes of the eyes.

*Diplopia*, or *double vision*, can occur when gazing to the affected side.

*Loss of reflexes* (loss of the *accommodation reflex* and the *pupillary light reflex*).

## CN IV: THE TROCHLEAR NERVE

The trochlear nerve is responsible for the motor innervation to the *superior oblique* muscle of the eye. The trochlear nerve is considered the *smallest* cranial nerve, but runs the *longest* distance because it exits the brain stem on the opposite side of the other nerves. It then enters the *cavernous sinus* and enters the *orbit* through the *superior* orbital fissure. The trochlear nerve supplies one innervation: the *superior oblique* muscle.

> A trochlear nerve lesion will result in *vertical diplopia* and the inability to gaze *downward* and *laterally*. The double vision that results from a cut trochlear nerve is caused by the paralysis of the *superior oblique muscle.*

## CN V: THE TRIGEMINAL NERVE

The trigeminal nerve is the *largest* cranial nerve and offers *sensory* and *motor* fibers to parts of the face (Fig. 4-3). For example, the *sensory* fibers of the trigeminal nerve occur over a large part of the skin at the head and face, the nasal cavity and sinuses, and the oral mucous membranes. In addition, the *motor* fibers provide innervation to the *muscles of mastication*. The trigeminal *ganglion* has unipolar neurons. The nerve is also divided into the *ophthalmic* (upper or $V_1$), *maxillary* (middle or $V_2$), and *mandibular* (lower or $V_3$) divisions. The upper and middle divisions are purely *sensory*, and the lower is both *motor and sensory*.

The *ophthalmic division* is the *sensory* innervation of the forehead, a portion of the scalp, the upper eyelid, and the *ethmoidal sinuses*. This division of the trigeminal nerve enters through the *superior* orbital fissure into the orbit and further divides into the *frontal*, *nasociliary*, and *lacrimal* branches of $V_1$.

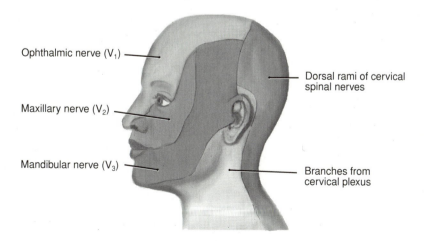

Ophthalmic nerve ($V_1$)

Maxillary nerve ($V_2$)

Mandibular nerve ($V_3$)

Dorsal rami of cervical spinal nerves

Branches from cervical plexus

**Figure 4-3**    Trigeminal nerve (CN V).

The *maxillary division* exits through the *foramen rotundum* and leaves the cranial cavity to the *pterygopalatine fossa*. This $V_2$ division of the trigeminal nerve acts as the *sensory* component of the skin covering the *maxilla*, nasal cavity, palate, and nasopharynx. In addition, it is the sensory component of the anterior and middle cranial fossa *meninges*. It further divides into the *meningeal, ganglionic, zygomatic* (which has a *zygomaticotemporal branch* to the lacrimal gland), *alveolar nerves*, and the *infraorbital nerves*.

### CN VI: THE ABDUCENS NERVE
The abducens nerve is responsible for the motor innervation to the *lateral rectus muscle* of the eye. It emerges from the brain at the pons-medulla junction and exits from the cranial cavity at the superior orbital fissure. Damage to the abducens nerve can result in *medial strabismus*, *diplopia*, and the inability of the eyes to move *laterally*.

### CN VII: THE FACIAL NERVE
The facial nerve is responsible for the motor innervation to the muscles of facial expression, the stapedius of the ear, as well as the stylohyoid and posterior belly of the digastric. In addition, it supplies several glands (lacrimal, oral, nasal, submandibular, and sublingual). The facial nerve is also responsible for the sensation of taste in the anterior two thirds of the tongue. In facial nerve paralysis, there is a *unilateral* loss of *taste* in the *anterior two thirds* of the tongue. In addition, there is decreased salivation and the lower face is affected (partial paralysis of the face). In *Bell's palsy*, the facial nerve is either infected or inflamed and this results in a paralysis of the facial muscles.

### CN VIII: THE VESTIBULOCOCHLEAR NERVE
The vestibular branch of this nerve is responsible for balance, and the cochlear portion is responsible for auditory sensation. The nerve enters the internal acoustic meatus (along with the facial nerve) and enters the foramina to supply the ampulla of the semicircular canals, the utricle, saccule, and the cochlea. As a result of damage to the vestibulocochlear nerve, the individual will experience *tinnitus* (ringing in the ears), *ataxia* (loss of equilibrium), some loss of hearing, and *vertigo* (imbalanced feeling of rotation). A common tumor of this area is the *accoustic neuroma*.

### CN IX: THE GLOSSOPHARYNGEAL NERVE
The glossopharyngeal nerve is responsible for general sensation from the pharyngeal mucosa, with the posterior one third of the tongue as well as the skin from behind the ear. In addition, the taste sensation is from the posterior one third of the tongue. The glossopharyngeal nerve has motor fibers to the stylopharyngeus, as well as fibers to the *parotid* gland. The *sensory* fibers to the tonsils also add to its functions. A lesion of the glossopharyngeal nerve results in pain in the throat and neck.

### ON X: THE VAGUS NERVE
The vagus nerve is involved in several functions of the physiologic system of the body. The vagus nerve supplies the motor innervation to the pharyngeal

muscles (except the stylopharyngeus), the palatal muscles (except the tensor palati), and the laryngeal muscles. In addition, it offers taste sensation from the root of the tongue and general sensation from the root of the tongue and behind the ear. Furthermore, the vagus nerve is an important part of the parasympathetic effect on the cardiovascular and other thoracic and abdominal organs.

The vagus nerve provides the following branches: the *pharyngeal*, the *superior laryngeal*, and the *inferior, or recurrent, laryngeal* nerves. A lesion of the vagus nerve can result in *tachycardia, decreased respiratory rate*, or gastrointestinal changes (vomiting). A deviation of the *uvula* toward the *intact* side may be noted. If the individual has damage to the *superior* laryngeal nerve, then there will be a weakened, monotonous, hoarse voice. The purpose of the "gag reflex" is to test the function of the vagus nerve and the response of the soft palate.

## CN XI: THE ACCESSORY NERVE

The accessory nerve is responsible for the motor innervation to the sternocleidomastoid (SCM) and trapezius muscles and distributes to the neck. The *spinal* accessory nerve enters the cranial cavity through the *foramen magnum* and exits through the *jugular foramen* with the glossopharyngeal and vagus nerves. Again, it supplies the *SCM* and the *trapezius* muscles. A damaged accessory nerve will result in "winging of the scapula." You can test for the SCM muscle by asking a patient to turn his or her head against the resistance of your hand. The same can be accomplished for the trapezius by asking him or her to move the shoulder against the resistance of your hand.

## CN XII: THE HYPOGLOSSAL NERVE

The hypoglossal nerve is responsible for the innervation to the *lingual* muscles (of the tongue), except the palatoglossus. Damage to the hypoglossal nerve results in a *unilateral paralysis* of the tongue. You will notice that the tongue deviates to the *ipsilateral* side of the lesion because the tongue is atrophied on the affected side, making it hard for the person to talk or eat.

# Notes

# Thorax

**5**

The chest wall (thoracic wall) is composed of the parietal pleura, the muscu-loskeletal layer (which includes the deep back muscles, superficial back mus-cles, innermost/internal/external intercostal muscles, and the ribs, sternum, and thoracic vertebrae), the superficial fascia, and the skin layer. The thorax protects the heart and lungs through its limited movement. There are 12 tho-racic vertebrae and 12 pairs of ribs (Fig. 5-1).

The "van," which represents vein, **artery**, and **nerve**, is the *neurovascular bundle*, which lies between the *internal* and *innermost* intercostal muscles. The van runs on the *upper* edge of the intercostal space (ICS) of the *lower* part of the rib. Therefore, when an intercostal nerve block is passed, it is done at the lower angle of the rib and at the upper ICS to block the nerve; always staying *lateral* to the angle of the rib. The sensory innervations of the intercostal nerves to the muscle, skin, parietal pleura, and the ribs allow for pain in this

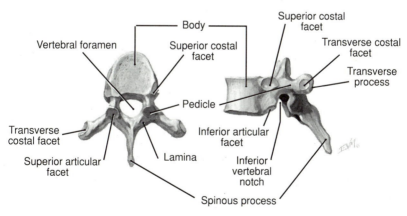

**Figure 5-1**  Thoracic vertebrae.

area (you can feel a broken rib). The pleural cavity is the fluid space that lies between the *visceral* and *parietal* pleura. A *pleural tap* enters through the skin and then passes through the following layers: superficial and deep fasciae, external/internal/innermost intercostal muscles, endothoracic fascia, *parietal* pleura, and the pleural cavity. (Note: Visceral pleura is fused to the *lung* itself, and it will NOT have pain with inflammation because there are NO somatic nerves in the pleura. The parietal pleura is fused to the inner chest wall. At the *hilus*, the two pleura are joined.)

If the *phrenic nerve* is damaged, you may expect half of the diaphragm to be paralyzed. This is best seen in an inspiratory radiograph, where the diaphragm is actually pushing upward instead of downward.

The **mediastinum** is the area *within* the thoracic cavity. It is divided into: superior, inferior, anterior, posterior, and lateral regions. The divisions are helpful for the different levels and regions. The regions of the mediastinum and its landmarks are summarized in the table that follows.

| Superior (Thoracic Inlet) | Inferior (Diaphragm) | Anterior (Sternum) | Posterior (Vertebrae) | Lateral (Mediastinal Pleura) |
|---|---|---|---|---|
| Esophagus, trachea, aortic arch and branches; superior vena cava and right and left brachio-cephalic veins and thymus | Diaphragm | Inferior portion of the thymus | Descending aorta, thoracic duct, azygos and hemiazygos veins; esophagus, vagus nerve, and sympathetic trunks | Pleura |

The *middle mediastinum* contains the pericardium and its contents (heart). It is important to recognize the landmarks in the thorax.

- **T2** Jugular Notch
- **T3** Top of Aortic Arch
- **T4** Sternal Angle
- **T7** Inferior Angle of the Scapula
- **T9** Xiphoid Process
- **T10** *Esophageal* Hiatus
- **T12** *Aortic* Hiatus

A cross section of the chest is shown in Fig. 5-2. Two types of sternal projections are reviewed in the following table.

| Pectus Excavatum | Pectus Carinatum |
|---|---|
| This is a "funnel chest," a rare congenital widening defect of the sternum body, which pushes against the heart and may cause an abnormality in the structure of the heart. The sternum is recessed inward. | This is also called "pigeon chest" and is characterized by the anterior projection of the sternum. |

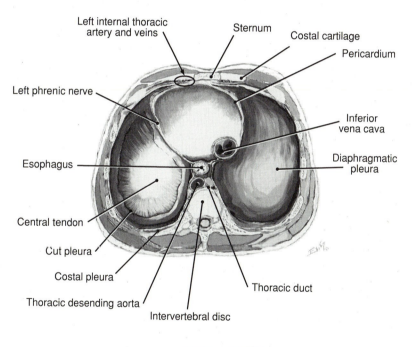

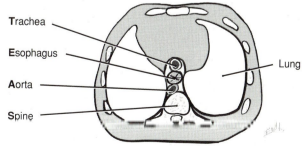

**Figure 5-2** "TEAS" cross section of chest.

## THE BREAST

The mammary glands, fat, and connective tissue create the majority of the breast. Each lobe contains a *lactiferous duct*, which exits at an opening at the nipple. The *lateral* quadrants, the nipple and the areolar area lymphatic system of the breast, drain to the *axillary* lymph nodes. The *medial* quadrants drain to the *parasternal* lymph nodes and to the *contralateral breast*. The breasts may be equal in size or at times asymmetrical. Most cancer develops in the highly *glandular* tissue of the *superolateral quadrant*. The vertebral *venous* plexuses are considered routes of spreading for cancerous cells of the breast to the *vertebrae* and then to the *skull* and *brain*. Breast carcinoma is spread via the *lymphatic* and *venous* systems. The cancerous cells may metastasize from the *lymphatic vessels* to the *axillary lymph nodes* and build nest formations within the nodes. The swollen nodes become hard and dimple the skin in breast cancer, creating the *peau d'orange*, or orange-peel skin, of the breast. There are different procedures involved in the removal of cancerous tissue of the breast (see the table that follows). Care must be taken in surgery to avoid cutting the *long thoracic nerve* which will paralyze the *serratus anterior muscle,* resulting in the condition called *winged scapula.*

*Polymastia*, or accessory breasts, can occur in certain individuals, and *polythelia*, or additional nipples, can occur along the line of embryonic mammaries. *Gynecomastia* can occur with alcoholism and is characteristic in some disorders. Klinefelter's syndrome, in which the individual carries an additional X chromosome (XXY), may be characterized by breast enlargement in males.

| | |
|---|---|
| **Mastectomy** | Removal (excision) of the breast |
| **Modified radical mastectomy** | Removal of the breast to deep fascia |
| **Radical mastectomy** | Removal of the breast, pectoral muscles, the fat and fascia, as well as the axillary and pectoral lymph nodes |
| **Lumpectomy** | Only the tumor and the surrounding tissue are removed |

## THORACIC SKELETAL AND NEUROMUSCULAR SYSTEM

There are **12** thoracic vertebrae posteriorly, a sternum anteriorly, with the ribs laterally in the thorax. The *sternum* consists of:

1. The *manubrium* (triangular bone, with a *jugular notch* at the **T2** level), the *manubriosternal joint* (at the *sternal angle of Louis* below **T4**), and the articulating surfaces of the *clavicle* and *ribs (numbers 1 and 2).*
2. The *body* below the manubriosternal joint and above the xiphosternal junction. The body is long and narrow and has articulating surfaces with *ribs (numbers 2 through 7).*
3. The *xiphoid process* is cartilaginous in the young and ossifies with age. It occurs at level **T9**.

The levels of T5 to T8 identify the thoracic aorta. In trisomy 21, there may only be 11 pairs of ribs (normally there are twelve pairs). The first 7 costal cartilages will likely always join the sternum, but you may encounter an individual with 6 or 8. In forked, or *bifid* ribs, there are usually 8 unilateral ribs. This is seen in less than 2 percent of the population. The *weakest* part of the rib is the *angle*, and it can bend and fracture under stress. The *clavicle* protects the initial two pairs of ribs, and the last two pairs are NOT joined (this allows them to be mobile, and they will not usually fracture). *Notching of the ribs* on x-ray is characteristic of *coarctation of the aorta*, where the engorged intercostal arteries form a "notch" in the *inferior* part of the rib.

The muscles of the thorax include the *intrinsic* muscles (external/internal/innermost intercostal muscles) and the *extrinsic* muscles, which join to the chest wall (trapezius, rhomboids, latissimus dorsi, and pectoralis major and minor). The neurovascular system (remember, "van" from superior to inferior) occurs in the intercostal spaces and provides innervation to the intercostal muscles, ribs, skin, and pleura.

## BREATHING

The *trachea* starts at level C6 and ends at the T4 level with a bifurcation to form the left and right main bronchi (Figs. 5-3 to 5-5). Remember, the *right* bronchus is *shorter*, *straighter*, and *wider* than the left main bronchus; there-

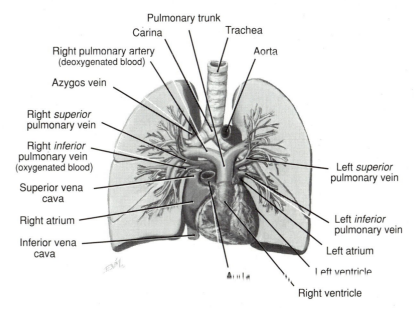

**Figure 5-3**  Pulmonary arteries and veins.

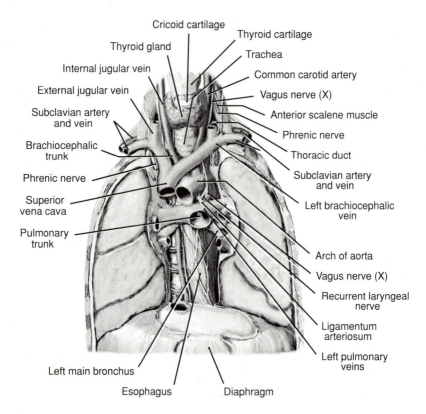

Cricoid cartilage
Thyroid cartilage
Thyroid gland
Trachea
Internal jugular vein
Common carotid artery
External jugular vein
Vagus nerve (X)
Subclavian artery and vein
Anterior scalene muscle
Phrenic nerve
Brachiocephalic trunk
Thoracic duct
Phrenic nerve
Subclavian artery and vein
Superior vena cava
Left brachiocephalic vein
Pulmonary trunk
Arch of aorta
Vagus nerve (X)
Recurrent laryngeal nerve
Ligamentum arteriosum
Left main bronchus
Left pulmonary veins
Esophagus
Diaphragm

**Figure 5-4**   Pulmonary arteries and veins.

fore, it is more likely that a foreign body (like a peanut) will enter the *right* bronchus. The *right* main bronchus divides into **3** *lobar* bronchi (these then divide into a total of **10** *segmental* bronchi). The *left* main bronchus divides into **2** *lobar* bronchi (which divide into **8** *segmental* bronchi).

Therefore, the *right* lung has **2** lobes separated by an *oblique fissure*; they are called the upper and lower lobes. The *left* lung has **3** lobes. The *upper* and *middle* lobes of the left lung are separated by the *horizontal fissure*. The *middle* and *lower* lobes are separated by the *oblique fissure*. Lymphatic drainage goes to the *pulmonary* nodes in the lung and then to the *bronchopulmonary* nodes at the *hilus*, followed by the *tracheobronchial* nodes, and finally, to the *thoracic duct* at the *left* side and the *right lymphatic trunk* at the *right* side. The autonomic nervous system supplies *sympathetic* innervation to cause **vaso**constriction and parasympathetic innervation to cause **broncho**constriction from the vagus nerve.

The *carina* is the ridge area between the right and left main bronchi. During a *bronchoscopy*, you may view the bronchi and this area. If there is

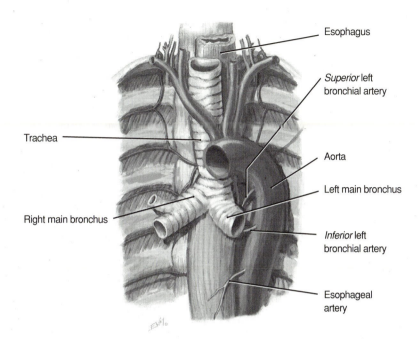

**Figure 5-5**   Bronchial arteries.

*bronchogenic carcinoma, lymphatic* spread can create enlargement of the lymph nodes and the displacement of the carina. The carina is very sensitive and will respond as the *cough reflex* at the mucous membrane. If a portion of food (like the common peanut) is passed through the bronchi, at the moment it passes the carina, it will not be able to be "coughed up" because the reflex response will be lacking. *Atelectasis* will occur *distal* to the lodged peanut (the lung collapses) and the person will experience *dyspnea*, or difficulty breathing. The usual site for a foreign body to enter is the *right* bronchus because it is *shorter, wider,* and *more vertical* than the left bronchus. This will then cause the object to enter the *middle* lobe or the *inferior* lobe.

The *base* of the lung is actually the inferior limit of the *posterior* surface of the *lower* lobe of the lung (NOT at the diaphragm area). It is necessary to listen to the lung at the 10th thoracic vertebra.

A list of several pleural cavity conditions is summarized in the following table. You should also recognize that a *stillborn* infant will not float because his or her lungs have not expanded. However, a live born infant that is found dead will float because the lungs were filled with air. Furthermore, you can recognize that the lungs are filled by percussion and auscultation.

| Hydrothorax | This is the accumulation of *fluid* in the pleural cavity. |
|---|---|
| Pneumothorax | This occurs as *air* entry into the pleural cavity following a penetration to the lung (i.e., ruptured lung or rib fracture). This may collapse a portion of the lung and may not effect the other *separate* lung sacs. *Positive*-pressure pneumothorax is considered a medical emergency because of the increased air in the pleural cavity that shifts the mediastinum to the opposite side—compressing the opposite lung. |
| Hemothorax | The accumulation of *blood* into the pleural cavity. This can occur following a chest wound. |
| Pleuritis | This is the *inflammation* of the *pleura* and creates referred pain to the intercostal nerve *cutaneous* (skin) distribution. |
| Friction rub | This is also called *pleural rub* and is associated with pleuritis creating a rough surface in the lung. It is audible with a stethoscope and may create referred pain to the thoracoabdominal wall or to the shoulder. |
| Pleural exudate | This is the accumulation of *serum* from the blood vessels into the pleural cavity. |

## PULMONARY THROMBOEMBOLISM

A *thrombus*, or blood clot, can detach and become an *embolus* passing through the *right* side of the heart. This flows through the *pulmonary **artery*** and enters the lung. This creates an *obstruction* of the arterial *blood flow* to the lung. The result is a pulmonary thromboembolism and possibly illness and death (morbidity and mortality). The obstruction of the pulmonary artery by an embolus will create an area of the lung that is NOT perfused—although it is ventilated. A *large* embolus at the pulmonary trunk creates *acute respiratory distress* followed by death. A *medium*-sized embolus can block the artery and create an *infarct* with necrotic tissue. In a healthy person, the body compensates for this loss by forming *collateral circulation* to supply the tissue and avoid an infarct.

## BRONCHOGENIC CARCINOMA

This is considered the most common cancer in men (it accounts for almost one third of the malignancies). Bronchogenic *carcinoma* is caused by carcinogens like cigarette smoking and metastasis via *lymphatics* (remember, *carcinomas* usually metastasize via the *lymphatic* route, and *sarcomas* metastasize via *hematogenous spread*). The cancer cells can also travel through the blood to the brain, lungs, bones, and the *supraclavicular* lymph nodes. On examination, the individual will likely have enlarged and hardened *supraclavicular lymph nodes* and may have malignancy involvement in other organs. The bronchogenic carcinoma may involve the tracheal and tracheobronchial lymph nodes since they drain the lungs. Clinically, it may affect the *phrenic nerve* and result in paralysis of half of the diaphragm. In the case of *apical lung cancer*—or with bronchial or esophageal cancers—the carcinoma can affect the *recurrent laryngeal nerve*. The recurrent laryngeal nerves supply all of the intrinsic muscles of the larynx. The disease of the *superior mediastinum* or a surgical laceration can result in *vocal cord* paralysis and presents as *hoarseness* in the person's voice. This nerve can also be *stretched* secondary to a dilated aorta (as seen in an aneurysm of the aorta).

# Notes

# *The Heart*

6

---

## THE HEART

The heart is a fist-sized muscular pump with two *atria* and two *ventricles* as chambers. The cardiac conduction system is made of *cardiac muscle cells*. The sinoatrial (SA) node is in the subepicardium at the *right atrium* near the superior vena cava (SVC). The SA node is considered the *pacemaker* of the heart because of the rapid depolarization of the cardiac muscle cells. Remember, these are NOT *nerve* cells, they are cardiac *muscle* cells. The depolarization "spreads" into the atrial muscular walls. The atrioventricular (AV) node occurs in the interatrial septum between the two atria and within the subendocardium. From here, the AV node passes the conduction to the AV *bundle*. These fibers conduct from the atria to the *ventricles*. Finally, the left and right bundle branches flow through the sides of the interventricular septum to the ventricles (Fig. 6-1).

---

## CIRCULATION OF THE HEART

The two *coronary arteries* from the *ascending aorta* supply the blood to the heart. The *right* coronary artery, with an anastomosis with the circumflex branch of the *left* coronary artery, provides the circulation to the *right* atrium and ventricle, the septum between the atria, and the SA and AV nodes. The *posterior interventricular (IV) artery*, with an anastomosis with the *anterior* IV artery, supplies the blood to the right and left ventricles. Another branch of the *right* coronary artery is the *right* marginal artery. This supplies the *right* ventricle. The *left* coronary artery comes from the aorta and travels behind the pulmonary trunk to the coronary sulcus and divides into the *anterior* IV artery and the *circumflex* artery. The *anterior* IV artery forms an anastomosis with the *posterior* IV artery to supply the *right* and *left* ventricles and the IV septum. The circumflex artery supplies the *left* atrium and ventricle and includes

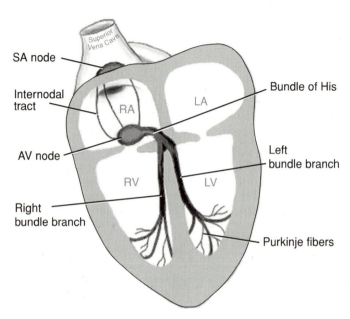

**Figure 6-1**   Conduction system of the heart.

an anastomosis with the *right* coronary artery. Heart circulation is illustrated in Fig. 6-2.

On physical examination of a child, you should palpate for the *apex beat* at the *fourth* intercostal space along the mid-clavicular line. This would be a space above the location for palpation (at the *fifth* intercostal space) in any toddler or older person. By *percussion*, you can tap your middle finger and recognize the changes in sound when moving from the lung to the heart. Notice that when you tap near the edge of the heart, there is a dull "thump" rather than the "hollow tap" of the lung space. This will give you an idea of how large and what shape the heart is. In a *hypertrophied* (enlarged) heart, consider a disease or problematic state (shunt, dilated ventricles, diseased valves, etc.).

A right *catheterization* is the passing of a catheter through the *external iliac vein* into the *inferior vena cava* (IVC) and through the *right* chambers or the pulmonary *artery*. During a left *catheterization*, the catheter is passed through the *femoral artery* and through the *aorta* to the *left* ventricle.

## RIGHT ATRIUM

The right atrium receives venous blood from the SVC, IVC, and the coronary sinus (which receives blood from the cardiac veins). In an *atrial septal defect*, oxygenated blood from the lungs is shunted from a defect in the *left* atrium and enters into the *right* atrium. What results is an *enlarged right* atrium and ventricle. Furthermore, the *pulmonary trunk* will dilate.

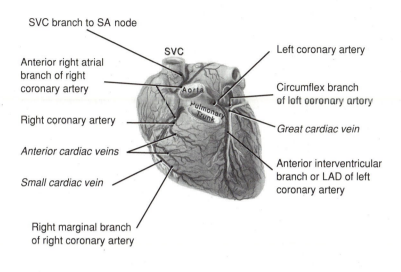

SVC branch to SA node

Anterior right atrial branch of right coronary artery

Right coronary artery

*Anterior cardiac veins*

*Small cardiac vein*

Right marginal branch of right coronary artery

SVC

Aorta

Pulmonary Trunk

Left coronary artery

Circumflex branch of left coronary artery

*Great cardiac vein*

Anterior interventricular branch or LAD of left coronary artery

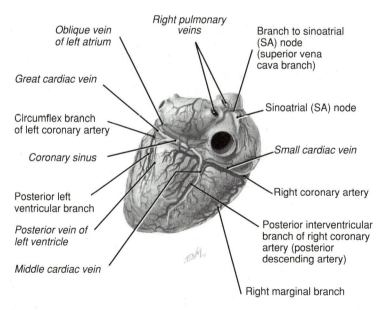

*Oblique vein of left atrium*

*Right pulmonary veins*

Branch to sinoatrial (SA) node (superior vena cava branch)

*Great cardiac vein*

Circumflex branch of left coronary artery

*Coronary sinus*

Posterior left ventricular branch

*Posterior vein of left ventricle*

*Middle cardiac vein*

Sinoatrial (SA) node

*Small cardiac vein*

Right coronary artery

Posterior interventricular branch of right coronary artery (posterior descending artery)

Right marginal branch

**Figure 6-2**   Coronary arteries and cardiac veins.

49

The venous drainage of the heart flows into the *coronary sinus* and then into the *right atrium*.

## RIGHT VENTRICLE

The right ventricle *receives* blood from the right atrium. The *pulmonary valve* has *three* semilunar cusps, which close and stop the back flow of blood. There are *three papillary muscles* in the right ventricle that attach with the ventricular wall (at the base) and the chordae tendineae (attached to the cusps of the right *atrioventricular valve*). The right AV valve is considered the *tricuspid valve*. In *pulmonary valve stenosis*, where the cusps of the valve fuse and narrow the opening, the right ventricle is likely to be enlarged (hypertrophy).

## LEFT ATRIUM

The left atrium has *four* pulmonary veins entering in the posterior wall, and this atrium is slightly thicker walled than the *right* atrium. Oxygenated blood travels from the left *atrium* into the left *ventricle*. Try to remember that in thrombi from the *left* atrial walls, which break off and embolize, the thrombi go into the *systemic* circulation and cause obstruction of the *peripheral arteries*. In the event that a cerebral artery is blocked, a *stroke* is caused by the disruption of an area of the *brain*—which supplies the control to an area of the *body*.

## LEFT VENTRICLE

The left ventricle is the *apex of the heart* and is the work area of the heart since it pumps the *systemic* circulation. The *left* ventricular wall is nearly twice as thick as the *right* ventricular wall, and the left ventricle connects to the outward flow of the *aorta*. The *left atrioventricular valve* is called the *mitral valve* and has *two* cusps. The *mitral* valve is the most commonly diseased heart valve. Scarring or nodule formation causes difficulty in blood flow. In addition, the usual *heart murmur* is commonly due to the mitral valve regurgitating blood into the left *atrium* while the left *ventricle* contracts. *Aortic valve stenosis* also causes the heart to work against more pressure and results in *left ventricular hypertrophy*. Because there is so much pressure ascending out of the heart, if the aorta is weak (as in Marfan's syndrome), an *aortic aneurysm* can result (the force of the left ventricle pushes the blood through and causes rupture in the aorta). Filling and pumping ventricles are shown in Fig. 6-3.

## VENOUS DRAINAGE OF THE HEART

The heart has veins that empty into the *coronary sinus* (main vein) of the heart, and small veins that empty into the *right atrium* (see Fig. 6-2 on coronary venous drainage and recognize the structures neighboring the vessels).

---

## PERICARDIUM

The external surface of the heart contains the *visceral layer* of the *serous pericardium* fused to the heart. There is a *pericardial cavity* between the *visceral* and *parietal layers*.

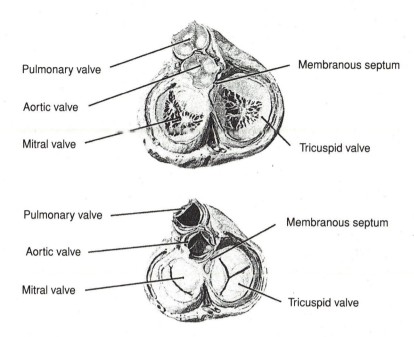

Pulmonary valve

Aortic valve

Mitral valve

Membranous septum

Tricuspid valve

Pulmonary valve

Aortic valve

Mitral valve

Membranous septum

Tricuspid valve

**Figure 6-3**    *Top:* Filling ventricles (diastole) (opened mitral and tricuspid valves). *Bottom:* Pumping ventricles (systole) (opened pulmonary and aortic valves).

## CLINICAL RELATIONS

### TETRALOGY OF FALLOT

*Tetralogy of Fallot* is considered the most common *cyanotic* congenital defect. It occurs as a misalignment of the *septum*—between the aorta and the pulmonary trunk—and the *ventricular septum*. As a result, the displaced septum causes *pulmonary stenosis* and an *overriding aorta* (enlarged outflow from the aorta that "overrides" the right and left ventricles). In addition, the displaced septum cannot fuse with the ventricular septum, creating a *ventricular septal defect*. Finally, pulmonary stenosis creates *right ventricular hypertrophy*. Thus, the *tetralogy* includes: *pulmonary stenosis, overriding aorta, ventricular septal defect*, and *right ventricular hypertrophy*.

### AV FISTULA

An *arteriovenous (AV) fistula* may be present in a patient with palpitations (tachycardia), fatigue, and difficulty with walking.

## CARDIAC TAMPONADE

*Kussmaul's sign* is seen when the neck veins bulge during *inspiration* from an increase in venous pressure.

## ATRIAL SEPTAL DEFECT

An *atrial* septal defect (ASD) can present in a patient as problems breathing and palpitations. The *foramen ovale* ASD location midseptal defect is the most common type of ASD. Overall, ASDs are the most common type of *congenital heart disease* in adults. *Atrial* septal defect (ASD) occurs in as many as one in four individuals between the left and right *atria* and will, with time, close after birth. They are associated with infective endocarditis and congestive heart failure (CHF). In an ASD, *oxygenated* blood from the *left atrium* is shunted into the *right atrium* and causes the *pulmonary system* to overwork. Thus, the *right* atrium and ventricle enlarge and the pulmonary trunk dilates. As a result of an *atrial septal defect*, pulmonary hypertension and *right*-sided heart failure may occur.

## VENTRICULAR SEPTAL DEFECT

*Ventricular* septal defect (VSD) is the most common *congenital heart defect*. The VSD occurs at the *interventricular septum* (the membrane between the two ventricles). The defect results in a *left*-to-*right* shunting of the oxygenated blood into the pulmonary system. This creates a *pulmonary hypertension*.

## PATENT DUCTUS ARTERIOSUS

Patent ductus arteriosus (PDA) is an *opened* "ductus arteriosus." The ductus arteriosus *failed to close* following birth. After birth, the embryonic flow (normally, pulmonary artery to aorta) reverses from aorta *to* pulmonary artery because of the reversed pressure gradient.

## COARCTATION OF THE AORTA

This congenital defect is a *purse-string constriction* of the *aorta*. Collateral circulation will develop if the coarctation occurs *inferior to* the ligamentum arteriosum. With an increased blood pressure, an individual may present on physical examination with the following: *increased* blood pressure at the *upper* extremities and *decreased* blood pressure in the *lower* extremities, and with a SEM (systolic ejection murmur). On further examination, an x-ray may reveal *rib notching* caused by collateral blood circulation, as well as a dilated aorta. The *narrowed* aorta, or *coarctation* of the aorta, usually occurs *distal* to the left *subclavian artery*. Turner's syndrome is also associated with coarctation of the aorta. The eventual hypertension can be corrected with surgery.

## AORTIC VALVE STENOSIS

The aortic valve may become *fused* at the margins and *narrow* the opening. This may be the result of a *congenital defect* or it may be *acquired*. The *left* ventricle workload is increased as a result of the stenotic valve and creates *left ventricular hypertrophy*.

## CEREBROVASCULAR ACCIDENTS (CVA)

Cerebrovascular accident (CVA), or a "stroke," can occur following an occluded *artery* in the *brain*. The result is paralysis of the motor innervations related to the affected area of the brain. To recognize which area of the brain may be affected, an understanding of the *homunculus* is expected. In other words, if the stroke victim experiences a paralysis in the right arm, then the CVA likely occurred at the area of the brain that controls the movement of the right arm.

## ANGINA PECTORIS

*Angina pectoris* presents as a *substernal* "chest pain" that may result from stress, myocardial ischemia, *arterial* constriction, or other causes. The *squeezing* or *tight* feeling is usually relieved following a couple of minutes of "rest." Arterial *stenosis* is the *narrowing* of the arterial flow. Arterial *atresia* is the *blockage* of the arterial flow.

In a *myocardial ischemic attack*, the pain will not decrease after a couple minutes of "rest." The myocardium is *ischemic*, and this lack of blood flow (infarct) results in a "heart attack." The heart muscle is damaged. The muscle will eventually be replaced with *fibrous tissue* and scar formation. Coronary *angiography* is the placement of a catheter through the *femoral artery* or the *brachial artery*, entering the *ascending aorta* to the *coronary artery*. The most common site is the LAD, or left anterior descending (the *anterior interventricular branch* of the left coronary artery). The second is the *right* coronary artery and then the *circumflex branch* of the left coronary artery. In surgery, a *coronary artery bypass graft (CABG, "cabbage")* provides blood flow from the *aorta* or the coronary artery to a branch of the coronary artery distal to the infarct.

Medications like *sublingual nitroglycerin* are helpful in *dilating* the *coronary arteries*, which increases the blood flow to the heart. This also relieves the pain.

Cardiac *referred pain* usually radiates from the *substernal* area or the *left pectoral* area to the *left* shoulder and arm (but it may also occur in the right or both arms). Pain is associated with angina and a myocardial infarction. It is due to the release of metabolic products and ischemia—which involves the afferent pain fibers.

## PERICARDITIS

The *pericardium* is the dual-walled sac that encapsulates the heart. Pericarditis is the *inflammation* of the pericardium and results in the *substernal pain*. The pulmonary veins can be compressed by pericardial *effusion*. This creates *cardiac tamponade*. The sign of *pericarditis* is pericardial friction rub (lean the patient forward and auscultate a forced expiration). By performing a *pericardiocentesis* (needle aspiration at the left fifth or sixth intercostal space), the additional fluid is drained from pericardial cavity to relieve the pressure. Be careful to avoid the *internal thoracic artery*.

## THE THYMUS GLAND

As a child grows, and up until puberty, the thymus gland *enlarges*. Then, in adulthood, the thymus becomes exchanged for fatty and fibrous tissue. *Thymomas* are benign tumors of the thymus and result in *retrosternal pain*, *coughing*, and difficulty breathing. This occurs because the thymoma compresses the neighboring structures like the *superior* vena cava, the *trachea*, and *veins* in the neck area.

# Notes

# Abdomen

## 7

## ABDOMEN

The abdomen is located between the *thorax* and the *pelvis*. The abdominal cavity is separated by the *diaphragm* and is continuous with the *pelvic cavity*. This cavity contains the abdominal digestive organs and is divided into *nine* regions. Take a careful look at the abdominal cavity and the included digestive organs. The abdomen contains the stomach, small and large intestines, spleen, liver, pancreas, kidneys, and glands. The anterior abdominal wall is composed of the following layers: skin, superficial fasciae (Camper's fascia, or the outer fatty layer; and Scarpa's fascia, or the inner fibrous layer), external oblique layer, internal oblique layer, transversus abdominis layer, transversalis fascia, extraperitoneal fatty layer, and the parietal peritoneum. The nine regions are also divided into *four* quadrants. Take the time to learn the different regions and organs shown in Figs. 7-1 through 7-2.

## VASCULATURE OF THE ABDOMEN

The abdomen is supplied by the *lower intercostal, lumbar, superior* and *inferior epigastric*, and the *inferior phrenic* arteries. In addition, the *middle suprarenal, renal*, and the *gonadal* arteries (all from the aorta) provide additional vasculature. The *celiac trunk* and the *superior* and *inferior mesenteric arteries* also supply the abdominal viscera (Figs. 7-3 through 7-8).

## MUSCLES OF THE ABDOMEN

The abdomen has *four* important muscles: the *external* and *internal obliques*, the *transversus abdominis*, and the *rectus abdominis*. An *acute abdomen* is described as an intense *pain* with spasm of the anterolateral abdominal muscles. *Guarding* or *splinting* is a palpable spasm of the abdominal muscles (the patient will not relax his or her abdominal muscles).

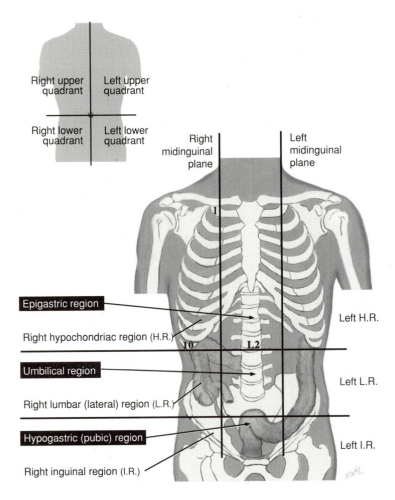

Figure 7-1    Regions of the abdomen.

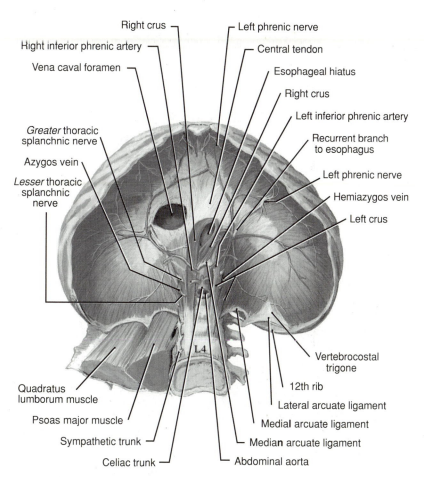

Right crus —

Right inferior phrenic artery —

Vena caval foramen —

*Greater* thoracic
splanchnic nerve

Azygos vein

*Lesser* thoracic
splanchnic
nerve

Left phrenic nerve

Central tendon

Esophageal hiatus

Right crus

Left inferior phrenic artery

Recurrent branch
to esophagus

Left phrenic nerve

Hemiazygos vein

Left crus

L4

Vertebrocostal
trigone

12th rib

Lateral arcuate ligament

Medial arcuate ligament

Median arcuate ligament

Abdominal aorta

Quadratus
lumborum muscle

Psoas major muscle

Sympathetic trunk

Celiac trunk

**Figure 7-2**  Abdominal diaphragm.

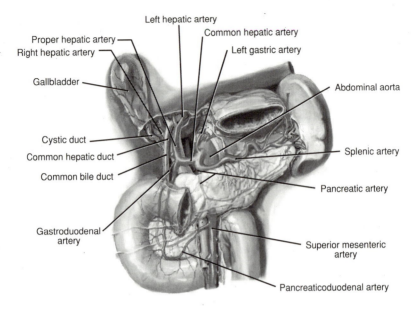

**Figure 7-3** Abdominal diaphragm.

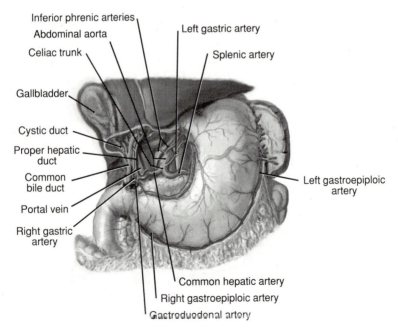

**Figure 7-4** Arteries near stomach.

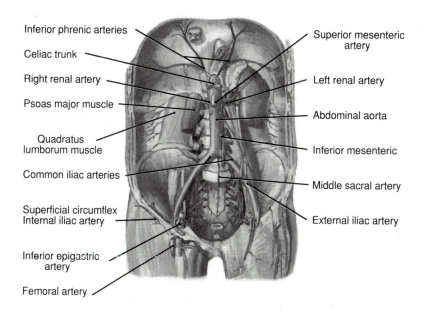

Inferior phrenic arteries

Celiac trunk

Right renal artery

Psoas major muscle

Quadratus
lumborum muscle

Common iliac arteries

Superficial circumflex
Internal iliac artery

Inferior epigastric
artery

Femoral artery

Superior mesenteric
artery

Left renal artery

Abdominal aorta

Inferior mesenteric

Middle sacral artery

External iliac artery

**Figure 7-5**   Arteries of posterior abdominal wall.

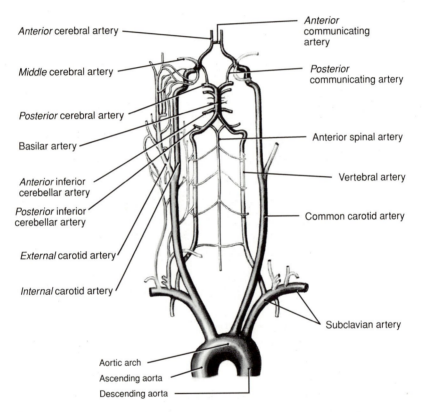

Anterior cerebral artery

Middle cerebral artery

Posterior cerebral artery

Basilar artery

Anterior inferior
cerebellar artery

Posterior inferior
cerebellar artery

External carotid artery

Internal carotid artery

Anterior
communicating
artery

Posterior
communicating artery

Anterior spinal artery

Vertebral artery

Common carotid artery

Subclavian artery

Aortic arch

Ascending aorta

Descending aorta

**Figure 7-6**  Arteries to the brain.

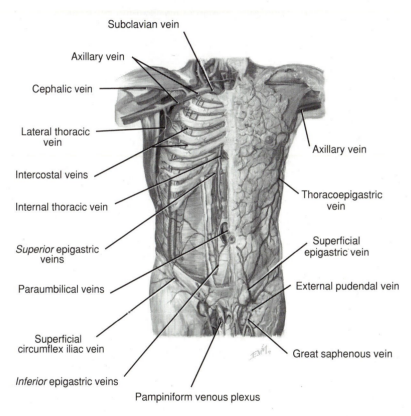

Subclavian vein

Axillary vein

Cephalic vein

Lateral thoracic vein

Intercostal veins

Internal thoracic vein

*Superior* epigastric veins

Paraumbilical veins

Superficial circumflex iliac vein

*Inferior* epigastric veins

Axillary vein

Thoracoepigastric vein

Superficial epigastric vein

External pudendal vein

Great saphenous vein

Pampiniform venous plexus

**Figure 7-7** Veins of anterior abdominal wall.

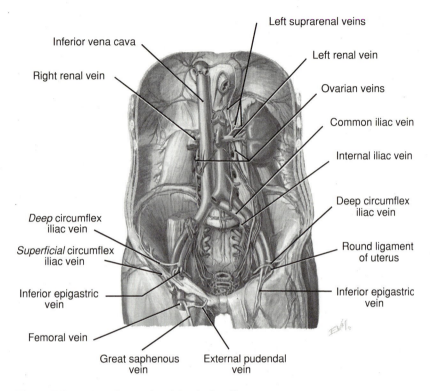

**Figure 7-8**   Veins of posterior abdominal wall.

## APPENDICITIS

Inflammation of the appendix, or *appendicitis,* usually presents as pain in the periumbilical region with nausea and vomiting. The pain then moves to the *right lower quadrant* of the abdomen. A gridiron or muscle-splitting incision is considered during surgical assessment and an appendectomy.

## HERNIAS

A *hernia* is a protrusion of a structure (like a visceral organ) from the cavity through a weakened area of the abdominal wall. Hernias are common in males and include: *indirect inguinal* and *direct inguinal* hernias.

### INDIRECT INGUINAL HERNIA

This is the most common type and is usually the *small intestine* that protrudes through the anterior abdominal wall (just *lateral* to the *inferior* epigastric vessels) and is covered by every layer of the spermatic cord. Indirect inguinal hernias are either congenital (birth) or acquired. Indirect inguinal hernias follow the descending testis through the inguinal canal; the hernia passes through the *deep* inguinal ring and exits at the *superficial* inguinal ring.

## DIRECT INGUINAL HERNIA

The direction of the hernia is *anterior* through the posterior wall of the inguinal canal. It leaves *medial* to the *inferior* epigastric vessels and enters the inguinal triangle. The *inguinal triangle* is posterolateral to the *superficial* inguinal ring. Direct inguinal hernias occur following a *weakening* of the anterior abdominal muscles from increased pressure within, such as weight lifting, pushing, and intense coughing. *Direct* inguinal hernias tear through the weakened abdominal wall. They usually reach the *superficial* inguinal ring (but unlike the *indirect* hernia, they do not enter the inguinal canal through the *deep* inguinal ring). This area is usually referred to as *Hesselbach's triangle*, or the *inguinal triangle*. The area is bounded by the *inferior epigastric artery*, the *inguinal ligament*, and the *lateral* border of the *rectus abdominis*.

## ACQUIRED HIATAL HERNIA

This can occur with increased pressure in the *abdominal* cavity (versus the thorax) and is usually found in middle-aged individuals. The hernia is designated as either a *sliding hiatus hernia* or a *paraesophageal hiatus hernia*.

---
## GENITAL AREA

Nerve supply to the *anterior third* of the *scrotum* is supplied by the ilioinguinal and genitofemoral nerves originating at the **L1** level. Nerve supply to the *posterior third* of the scrotum is from the perineal and posterior femoral cutaneous nerves. Testicular tumors tend to spread via the *lymphatic* system to the *lumbar lymph nodes*. With cancer of the scrotum, the spread may penetrate the *superficial inguinal* lymph nodes. The procedure for sterilization in a man is called *deferentectomy* or *vasectomy* and is usually done by bilateral ligation of the *ductus deferens* in an attempt to stop the passage of sperm through the urethra. This will NOT affect the auxiliary glands' secretions or ejaculation.

---
## HYDROCELE AND HEMATOCELE

When there is excess *fluid* in the processus vaginalis, this is considered a *hydrocele*. A *hematocele* is a *blood*-filled cavity in the tunica vaginalis following trauma to the scrotum or testes.

---
## DIAPHRAGM

The diaphragm is made of a *muscular portion* and a *central tendon*. The diaphragm is innervated by the *phrenic nerves*. Remember the following apertures.

1. *Caval hiatus* is at the level of **T8**. It allows the inferior vena cava and a few branches of the right phrenic nerve.
2. *Esophageal hiatus* is at the level of **T10**. This opening allows the esophagus and the anterior and posterior vagus nerves to pass through. It lies within the muscle of the right crus to the left of the midline.
3. *Aortic hiatus* is at the level of **T12**. This allows the aorta, azygos vein, and the thoracic duct.

## *MESENTERIES*

The mesenteries are bilayered sheets of peritoneum which connect the *parietal* and *visceral* peritoneum and cover the peritoneal organ. The following mesenteries are two examples of how mesenteries are given specific names because of the associated organs.

*Falciform ligament*—connects the liver to the parietal peritoneum of the anterior wall.
*Lesser omentum*—connects the liver to the *stomach* and *duodenum*.

# Notes

# Pelvis and Perineum

---

## 8

## PELVIC APERTURE

The female pelvic *inlet* is measured during a gynecologic examination, and the anterior-posterior diameter is measured from the midpoint of the *superior pubic symphysis* to the *sacral promontory*. The greatest width of the pelvic aperture is the *transverse diameter*. This is from the *linea terminalis* and goes from one side to the other. The *oblique diameter* is measured from the iliopubic eminence to the sacroiliac joint. The *birth canal* provides the path for the fetal head (and then body) to exit. The drug *relaxin* causes the relaxation of the pubic symphysis. The *pubococcygeus muscle* is the muscle most likely to be damaged during the birth of a child. An *episiotomy* will ease the delivery process, but the physician must be careful not to damage the *puborectalis* muscle. A *cystocele* is a herniated bladder that enters into the vagina and can be seen if the *pelvic fascia* or the *pubococcygeus muscle* is injured. The *cystourethrocele* involves herniation of the *urethra*.

Many elderly people may experience a fall, landing on the buttocks area. This fall may cause an injury to the *pubic rami,* the *acetabula,* and the *femor heads*.

---

## PELVIC VASCULATURE

The pelvis is supplied by the *common iliac arteria* (from the aorta), which is divided into the *external* and *internal iliac arteries*. The *internal iliac artery* is the primary blood supply to the pelvis, which divides into posterior and anterior divisions. The anterior division provides the main vessels to the *visceral* systems (uterine, umbilical, and rectal arteries). The *external* iliac artery does not really enter the pelvis but instead provides two arteries, the *deep circumflex iliac* and the *inferior epigastric*. When it passes under the inguinal ligament, it forms the *femoral artery*.

The *inferior vena cava* and the *portal vein* provide the venous drainage of the pelvic region.

## PELVIC INNERVATION

The pelvis and the viscera are innervated *autonomically* by the *sympathetic* and *parasympathetic* systems. The *splanchnic* nerves provide the *sympathetic* innervation, and the *vagus* (innervates the GI tract up to the splenic flexure) and *pelvic-splanchnic nerves* (innervate distal to the splenic flexure, the bladder, the uterus, prostate, and vagina) provide the *parasympathetic* innervation.

## UTERUS

The uterus may soften at the *isthmus,* and the cervix may feel separated from the body. This is the *Hegar's sign* and is an early sign of a pregnancy. Sperm usually fertilize the oocyte in the *ampulla* of the uterine tube. Once the new zygote becomes a *morula*, it travels into the uterus. Blockage of the *tubes* interferes with sperm's ability to travel and reach the oocyte. *Tubal ligation* is a technique used for birth control. In an infection or inflammation of the uterine tube, this is considered *salpingitis* and if there is *pus* present, *pyosalpinx*. An ectopic tubal pregnancy occurs if the zygote does not pass into the uterus. Instead, it implants into the uterine *tube* (most likely the *ampulla*). Unfortunately, a dangerous problem is the *ruptured tubal pregnancy*. This presents similar to acute appendicitis (the appendix is near the ovary). The uterus is illustrated in Fig. 8-1.

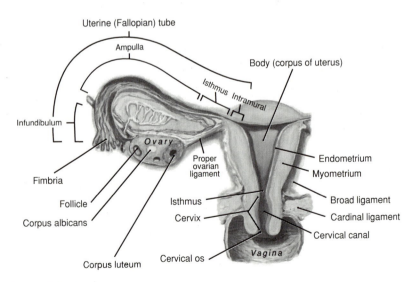

**Figure 8-1**   Uterus.

## HYMEN

The *hymen* is the inferior closed end of the uterovaginal canal and in infancy and childhood, this vaginal orifice is actually *closed*. Prior to puberty, it may rupture or perforate and menstrual fluid is released. During intercourse, the hymen will tear. It can also be a sign of child abuse.

## MALE GENITALIA

The male genitalia is composed of the testis, scrotum, gubernaculum testis, penis, penile urethra, and bulbourethral gland. The penis has three erectile tissues, two *corpora cavernosa* and one *corpus spongiosum*. When the arteries become vasodilated, through *parasympathetic* innervation of the erectile bodies of the arteries, *erection* occurs ("*p*" for "point"). Ejaculation of semen occurs after the contraction of the *bulbocavernosus muscle,* which is *sympathetic* innervation, and the perineal nerve ("*s*" for "shoot").

## FEMALE GENITALIA

The female genitalia is composed of the ovaries, labia majora and minora, clitoris, and the Bartholin's vestibular gland. The labia majora are the fatty folds of skin that outline the vagina. The thinner labia minora meet at the midline to form the *prepuce* of the *clitoris*. The clitoris is an *erectile* tissue body. The vagina enters through to the cervix of the uterus at the *fornix*.

# Notes

---

## VERTEBRAL COLUMN

This column has *vertebrae* and the *intervertebral disks* between the vertebrae (Fig. 9-1). The anterior part of the vertebra, which supports the weight of an individual, is called the *body*. The posterior part, which protects the spinal cord and meninges, is called the *neural arch*. The neural arch is made of the *pedicles* and the *laminae*. These surround the vertebral foramen. Together, the vertebral foramina create the *vertebral canal*.

There are **7** *cervical vertebrae*, and the first cervical vertebra is called the *atlas* (which does NOT have a body). The *second* cervical vertebra is the *axis*, which has the *dens* process from the body. The vertebral artery and vein will pass through the first **6** foramina in the transverse processes (called *transverse foramina*) of the cervical vertebrae. Finally, the vertebral artery enters through the *foramen magnum* (at the base of the skull).

There are **12** *thoracic vertebrae* where the *heads* of the ribs articulate with the *bodies* of the vertebrae—with the exception of rib numbers 1, 11, and 12, which articulate with only one body. The *tubercle* (of the rib) articulates with the *transverse process* of the vertebra.

There are **5** *lumbar vertebrae*, and these are considered the *largest* vertebrae with expansive space between the laminae. A *lumbar puncture* is conducted through these *interlaminar spaces* (between the lumbar vertebrae).

The final *sacral vertebrae* are *fused* and form the *sacrum*. The spinal nerves pass through the sacral foramina (there are no intervertebral foramina). A *caudal epidural block* is conducted through the caudal opening (at the end of the sacrum).

The *intervertebral disks* are made of an outer *anulus fibrosus*—made of fibrocartilage and connective tissue, and an inner *nucleus pulposus*—made of a semigelatinous, relatively acellular fluid.

The *intervertebral foramen* is bounded by the *pedicles* (superior and inferior), the *bodies* and the intervertebral disk (anterior), and the *articular processes* and *zygapophyseal joint* (posterior). The intervertebral foramen

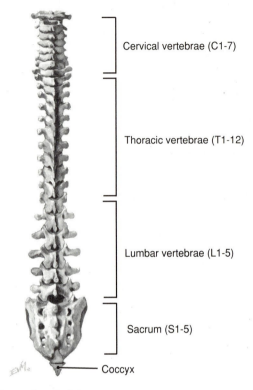

**Figure 9-1**   Vertebral column: dorsal view.

contains the dorsal root ganglion (DRG) and the spinal nerve as well as blood vessels, which supply the spinal cord.

The *vertebral ligaments* assist the vertebrae. The supraspinous, interspinous, and posterior longitudinal ligaments and the ligamentum flavum are the *posterior ligaments*, which resist *flexion*. The *anterior longitudinal ligament* resists *extension* of the vertebrae.

---

## BACK MUSCLES

The *superficial* back *intrinsic* muscles attach from the back to an upper limb. The *deep* back *extrinsic* muscles are deep in the thoracolumbar fascia and attach at both ends to the back or to the skull. The deep back muscles direct the movement of the back and resist the effect of gravity, allowing a person to stand erect. *Back strain* can occur with any excessive extension or stretch (even tears) of the muscle fibers and ligaments in the back. This includes the *erector spinae muscles*.

## SCIATICA

Muscle spasm of the lower back results in lower back pain with rigidity. The pain originates from lumbar disk protrusions onto the *sciatic nerve* in the L4-5 or L5-S1 locations.

## SPINAL NERVES AND THE SPINAL CORD

The *cell bodies* of the motor fibers are in the ventral horn of the spinal cord gray matter, and the cell bodies of the sensory fibers are in the *dorsal root ganglion*. Each *dorsal root* (which contains the *sensory* nerve fibers) and *ventral root* (*motor* fibers) join to form the **31 pairs** of **spinal nerves**. The spinal nerves divide into two *rami*, called the dorsal (posterior) and ventral (anterior) rami, which contain *both* motor and sensory fibers.

The *dorsal rami* provide branches to the motor and sensory innervation of the skin of the back, posterior scalp, deep back muscles, and the zygapophyseal joints of the vertebral column. The *ventral rami* provide other spinal innervated muscles (superficial back, skin and muscles of the lateral and anterior trunk, as well as the upper and lower limbs). The *cervical nerves* are numbered as they are in relation to the vertebra *below* the intervertebral foramen. The other nerves are numbered as the vertebra *above* the intervertebral foramen. The spinal cord segments are numbered the same as the nerve rootlets, which attach to the specific cord level.

## MENINGES

The three layers of the meninges (from inside out) are:

1. The **pia mater** is fused to the surface of the spinal cord.
2. The **arachnoid** is the layer above the pia mater. The *subarachnoid space* contains the *cerebrospinal fluid* and lies between the pia mater and the arachnoid.
3. The **dura mater** is the outer layer.

The *epidural space* lies outside of the dura and contains fat and a venous plexus.

## SPINAL TAP

A spinal tap is done below the spinal cord level and between the *subarachnoid space*. This is the level of L4 and just above the *iliac crest* at the level of the *cauda equina*. A *lumbar puncture* is performed between the *interlaminar space* on a flexed vertebral column. You will need to remember the layers that the needle passes through (it will NOT go through any back muscles since you are midline). These are:

1. Skin
2. Superficial fascia
3. Deep fascia
4. Supraspinous ligament
5. Interspinous ligament
6. Interlaminar space
7. Epidural space
8. Dura
9. Arachnoid
10. Subarachnoid space

A spinal block is a *direct injection* into the CSF and affects the nerve within a *minute* (a rapid effect); while an *epidural block* is injected into the *epidural space* and goes through the dura, arachnoid, and subarachnoid space with the CSF. In epidural block, the agent acts on the nerve roots and the spinal cord and this results in the loss of sensation *below* the level of the injection. The epidural block is used in decreasing the pain of parturition (birth) and its effect takes about 15 minutes to occur.

The role of the CSF is to prevent *spinal cord injury* during the body's continuous movements and in trauma. This fluid is easily accessible at the L3-4 or L4-5 levels (*lumbar puncture*).

## KYPHOSIS

The *humpback*, or posterior-convexed position of the vertebral column, creates a C-shaped (when viewed from the side) spine in the thoracic area.

## SCOLIOSIS

*Crooked back*, or a lateral curvature to the side, is the *scoliosis* of the vertebral column. More women have this than men, and it is the most commonly observed curvature. It may be corrected.

## LORDOSIS

This is the *backward-bent anterior* curvature, it is also known as *saddleback*. The *lower* back is sometimes affected in pregnant women, called *lumbar lordosis*. It is also associated with obesity and low back pain.

## SPINA BIFIDA OCCULTA

This congenital malformation of the vertebra results in *failure* to fuse the laminae of the vertebral arch at L5 through S1.

# Notes

# Lower Limb                                    *10*

---

## LUMBOSACRAL PLEXUS

The lumbosacral plexus is created from the *anterior rami* of T12 to L3 spinal nerves. The lower abdominal wall, perineal region, and the lower limb are innervated by branches of the lumbosacral plexus (**L2-S3**). The *femoral, obturator, tibial,* and *common peroneal nerves* are the major contributing nerves of the lower limb. The *femoral* nerve supplies the *anterior* area of the thigh, the *obturator* nerve supplies the *medial* area of the thigh, and the *tibial* nerve supplies the *posterior* area of the thigh (and leg). The common *peroneal* nerve innervates the *lateral* (superficial peroneal nerve) and *anterior* compartments of the leg (deep peroneal nerve).

---

## HIP JOINT

The hip is formed with the head of the femur and the acetabulum of the os coxae (parts of each of the *ilium, ischium,* and *pubis*). The capsule of the hip joint has three ligaments surrounding it: the *iliofemoral, ischiofemoral,* and *pubofemoral* ligaments. These deter the *extension* of the hip.

---

## KNEE JOINT

The knee joint is created by the *medial femoral, lateral femoral,* and *tibial condyles,* with the *patella.* The synovial cavity lies between the condyles and the patella and the femur. The *patellar ligament* attaches the patella to the tibia (anteriorly). The ligaments that resist hyperextension include the oblique popliteal and arcuate popliteal ligaments. The *medial* collateral ligament resists **ab**duction, and the *lateral* collateral ligament resists **add**uction. The anterior (ACL) and posterior cruciate ligaments (PCL) offer more stability (the ACL attaches to the tibia *anterior* to the posterior cruciate ligament; and because the ligaments cross, they attach on the opposite side of the femur).

The ACL and PCL prevent the anterior or posterior movement of the tibia on the femur. The *anterior drawer sign* relates to DAMAGED *anterior* cruciate ligament. In this case, the knee has *anterior slipping* of the tibia on the femur (the leg moves forward as the knee goes back). Another important ligament is the *medial meniscus*. This is a crescent-shaped ligament, which attaches to the tibia in front of the ACL and is attached to the medial collateral ligament. The *lateral meniscus* is O-shaped, also attached to the tibia (but not so strongly). The lateral meniscus is NOT attached to the lateral collateral ligament; instead, it attaches to the medial femoral condyle.

## ANKLE JOINT

The ankle bones have several joints. The *talocrural, subtalar*, and the *transverse tarsal* joints are three that you should be familiar with. Dorsiflexion and plantar flexion are allowed through the *talocrural* joint (between the distal tibia/fibula and the talus). In addition, the *medial collateral* or *deltoid ligament* is the "sprained" ligament seen in an *eversion* ankle injury. On the other side, the *lateral collateral ligament* is the "sprained" ligament seen in an *inversion* ankle injury. Another important ligament is the *plantar calcaneonavicular*, or *spring*, *ligament*, which is responsible for maintaining the arch of the foot.

## VEIN GRAFTS

In the case of a vein graft (to perform a bypass surgery for obstructed heart vessels), the *great saphenous vein* is delicately removed. After dissecting out the great saphenous vein, the physician will *reverse* the vessel in order to maintain the blood flow. Remember, there are several veins in the lower extremities and by removing this vessel, you will not cause difficulty in the return of blood from the limb to the heart.

## THROMBOPHLEBITIS

If a vein forms a *thrombus* (clot that is attached to the vessel), and the thrombus breaks off, it can travel through the *right* side of the heart and into the *lung* to form a *pulmonary embolus*. The large emboli may become entrapped and cause *pulmonary thromboembolism* or a sudden death syndrome in adults.

## COMMON INJURIES

### RUNNER'S KNEE

On occasion, you may come across an individual runner who complains of soreness and aching around the patella. This is usually caused by trauma (pull, blow, kneeling too much, flexing the knee too much, etc.) and the compression of the knee and movement of the patella. Runner's knee is a term for *chondromalacia patellae*.

## POPLITEAL CYST
The popliteal cyst is not really an "injury," but it does cause interference with the *movement* of the *knee*. It is also known as *Baker's cyst* with synovial fluid effusion.

## BURSITIS
*Prepatellar bursitis* occurs as *inflammation* of the knee as a result of the friction created between the individual's skin and the *patella*. It is also called *housemaid's knee* (workers, such as gardeners and carpenters, who may work on their knees without wearing kneepads).

## PULLED GROIN
When an individual overstretches or strains the *anterior* and *medial* muscles of the *thigh*, including the *iliopsoas muscle* and the **add***uctor muscle group*, this is considered a *pulled groin*.

## FOOT DROP
If the *common fibular nerve* is injured, the foot may *drop* as a result. In addition, injury to the *gluteal branches* of the *posterior femoral cutaneous nerve* can result in the loss of sensation of the *skin*. The common fibular nerve is the *most injured* nerve in the lower extremity. In clinical terms, the injury will cause *paralysis* of the *dorsiflexion* and *eversion* muscles of the foot, resulting in *foot drop* and some loss of sensation. These individuals tend to have a high step (to compensate for the drop) and to bring the foot down abruptly.

## FEMORAL NERVE
Injury to the femoral nerve results in the paralysis or paresis of the *anterior* thigh muscles or the *quadriceps femoris* muscles. The quadriceps femoris prevents the limb from going into full flexion. In essence, with such a lesion, the individual will have a "locked knee" in a fully extended position. Also, the patellar reflex is decreased.

## TIBIA
The tibia is the most commonly injured and fractured *long bone*. Such a fracture can occur following a skiing accident (torsion), a hard blow (compound fracture), or through disease like rickets. In rickets, a fracture is likely to occur at the *middle* and *inferior* thirds of the tibia. The common artery that is damaged is the *nutrient artery* because the fracture develops through the *nutrient canal*.

## TIBIAL NERVE INJURY
Although this nerve is NOT commonly injured since it is protected in the *popliteal fossa*, a deep laceration can cause injury and result in paralysis of the *flexor muscles* of the *leg*. In addition, it will affect the *intrinsic muscles*

of the sole of the affected foot (causing the inability to *plantarflex* the foot and toes). In addition, there will be some loss of sensation.

## FIBULA

A fractured fibula is commonly seen as a *lateral malleolus* fracture with an inverted foot. It usually is the result of slipping (on ice or other wet surface). Although the fibula does not regenerate, this bone is used as a *bone graft*. (This does NOT compensate the *function* of the leg.)

## SCIATIC NERVE

The sciatic nerve is the combination of the *tibial* and *common peroneal nerves.*

## INTERMITTENT CLAUDICATION

As a result of *arteriosclerosis* (stenotic arteries), the leg muscles receive decreased blood flow and the *ischemia* of the leg muscles creates *intermittent claudication*. The *posterior tibial artery* will not have a pulse. If you *palpate* the *dorsalis pedis* pulse, you will find that the *dorsum* of this foot will have a decreased (or absent) pulse.

# Notes

# Upper Limb                    *11*

---

## CLAVICLE

The clavicle contacts the *axial* skeleton and is a common location for fractures. The weakest areas of the clavicle are at the *medial two thirds* and the *lateral third* (medial to the *coracoclavicular ligament*). The clavicle is also the *first* bone in the body to ossify. This usually occurs in the seventh embryonic week as *intramembranous ossification*.

---

## HUMERUS, ROTATOR CUFF, SHOULDER MUSCLES, AND ELBOW

The upper extremity can begin with the *clavicle,* or collar bone, which has a sternal extremity that articulates with the sternum and an articular facet for the first rib. The *scapula* is a flat and triangular-shaped bone with a concave subscapular fossa, a posterior surface with other fossae, an acromion, or flattened area, with a scapular notch. The *clavicle* and the *scapula* comprise the *shoulder girdle*, and this joins at the sternoclavicular joint and to one another at the acromioclavicular joint. The shoulder joint resides between the *scapula* and the *head of the humerus.*

As you can see in Figs. 11-1 and 11-2, the *humerus* is the largest and longest bone in the upper extremity. Notice the surgical neck constriction below the tubercles; this is a frequent site of fracture. If you accidently injure the middle of the humerus, you may have injured your *radial* nerve.

The muscles of the shoulder begin with the *rounded* abilities of the *deltoid muscle* (Fig. 11-3). It is an important muscle to abduct the arm (humerus). But the deltoid muscle canNOT initiate this movement. It requires the assistance of the *supraspinatus* muscle. The shoulder joint is strengthened with the *supraspinatus*, *infraspinatus*, and the *teres major* tendons. Memorize the illustration in Fig. 11-4, as it contains the *shoulder muscles* and the *rotator cuff muscles.*

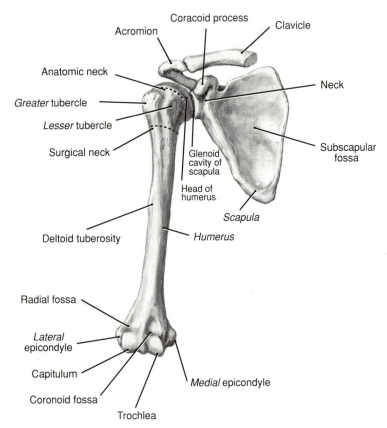

Coracoid process
Acromion
Clavicle
Anatomic neck
Neck
*Greater* tubercle
*Lesser* tubercle
Surgical neck
Glenoid cavity of scapula
Subscapular fossa
Head of humerus
Deltoid tuberosity
*Scapula*
*Humerus*
Radial fossa
*Lateral* epicondyle
Capitulum
Coronoid fossa
*Medial* epicondyle
Trochlea

**Figure 11-1** Anterior view of humerus and scapula.

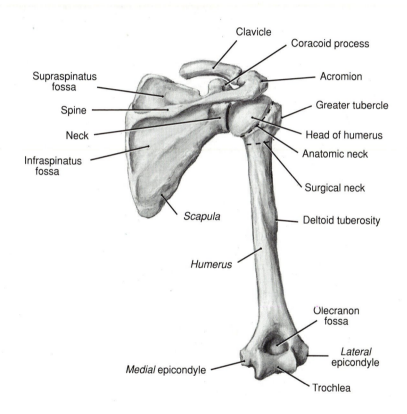

Clavicle

Coracoid process

Supraspinatus fossa

Acromion

Spine

Greater tubercle

Neck

Head of humerus

Infraspinatus fossa

Anatomic neck

Surgical neck

*Scapula*

Deltoid tuberosity

*Humerus*

Olecranon fossa

*Lateral* epicondyle

*Medial* epicondyle

Trochlea

**Figure 11-2**   Posterior view of humerus and scapula.

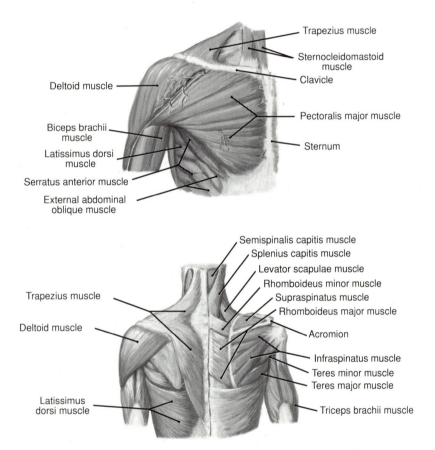

Trapezius muscle

Sternocleidomastoid muscle

Clavicle

Deltoid muscle

Pectoralis major muscle

Biceps brachii muscle

Latissimus dorsi muscle

Sternum

Serratus anterior muscle

External abdominal oblique muscle

Semispinalis capitis muscle

Splenius capitis muscle

Levator scapulae muscle

Rhomboideus minor muscle

Supraspinatus muscle

Rhomboideus major muscle

Trapezius muscle

Deltoid muscle

Acromion

Infraspinatus muscle

Teres minor muscle

Teres major muscle

Latissimus dorsi muscle

Triceps brachii muscle

**Figure 11-3**  Shoulder muscles: anterior view (top) and posterior view (bottom).

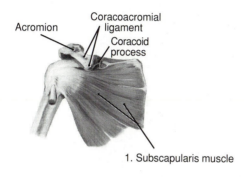

Acromion
Coracoacromial
ligament
Coracoid
process

1. Subscapularis muscle

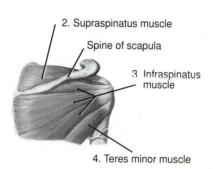

2. Supraspinatus muscle

Spine of scapula

3. Infraspinatus
muscle

4. Teres minor muscle

**Figure 11-4**  Rotator cuff muscles.

The *cubital fossa* is the triangular area in *front of* the elbow (Figs. 11-5 and 11-6). You should be aware of the *median cubital vein*. This is the great vessel that you access for intravenous blood samples, IV feeding, and other medical reasons. This vein lacks the large cutaneous nerves. It is a superficial vein and should be located with some ease (remember, *median is middle and a vein*).

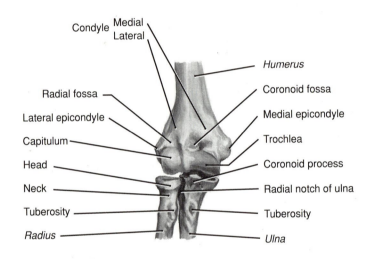

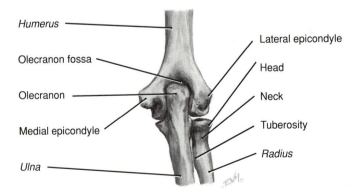

**Figure 11-5** Right elbow bones: anterior view (top) and posterior view (bottom).

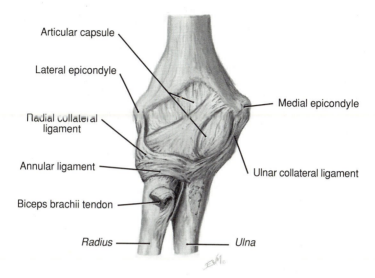

**Figure 11-6** Right elbow ligaments: anterior view.

## COMMON INJURIES

### LONG THORACIC NERVE INJURY

Injury of the long thoracic nerve results in paralysis of the *serratus anterior muscle*. This causes the scapula to fan outward, and the individual will not be able to *raise* or *push* the upper limb to the full extent. This injury (which occurs at **C5-7**) is also known as *winged scapula*. The serratus anterior muscle may also be injured by a stabbing or during surgery. This injury causes the nerve to become *compressed* and "wing" the scapula.

### UPPER BRACHIAL PLEXUS INJURY

This injury may be seen following a *football* or *motorcycle injury*—stretching the neck and the shoulder. The other movement that may cause an upper brachial plexus injury is the stretching or tearing of the plexus during delivery (pulled neck of the infant).

### LOWER BRACHIAL PLEXUS INJURY

This is actually not really a "common" injury, but it is important to be aware of. During a delivery, the physician may pull the infant too fast or too forcefully. This results in a decrease in motion and sensation in the wrist and fingers.

## AXILLARY NERVE INJURY

The *axillary nerve* travels around the *neck* of the *humerus* (at **C5-6**) and if it is injured, paralysis of the *deltoid muscle* and muscle atrophy will result (the shoulder will lose its "rounded" appearance). In order to test the deltoid muscle for atrophy or lack of strength, the examiner will **ab**duct the arm and then ask the patient to hold the arm in that position *against* a resistance. The injury also causes a loss of sensation on the *lateral proximal part* of the arm (Fig. 11-7).

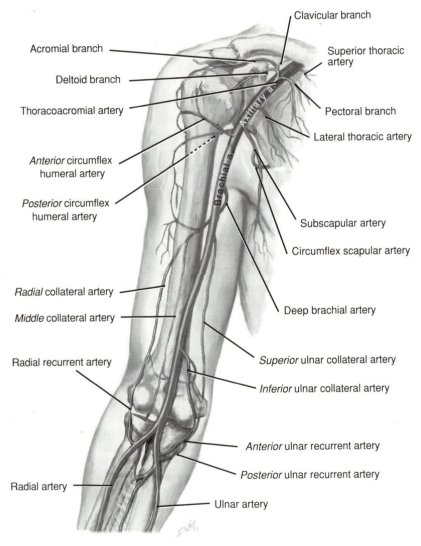

**Figure 11-7**   Axillary artery, brachial artery, and elbow anastomoses.

## THORACODORSAL NERVE INJURY

The *thoracodorsal nerve* (at **C6-8**) can be injured during surgery in the *axilla*. This can result in the paralysis of the *latissimus muscle* (Fig. 11-8).

## MEDIAN NERVE INJURY

Injury to the median nerve (proximal to the elbow) causes loss of *sensation* to the *lateral* part of the palm and digits. In addition, there is loss of motion in the thumb and in the *flexion* of the wrist and digits. If the median nerve is injured at the *elbow*, there is loss of flexion of the *proximal* interphalangeal joints. This may result following wrist-slashing in suicidal patients. If the median nerve is *compressed* near the elbow, pain may result in the anterior forearm and present with muscle hypertrophy. This is known as *pronator syndrome*. When the *median nerve* is compressed either by inflammation or arthritis, there is a loss of coordination and motion of the thumb and numbness of the digits. The location of carpal tunnel compression is at the *flexor retinaculum*.

## ULNAR NERVE INJURY

Injury to the ulnar nerve in the arm creates difficulty with movement of the *thumb, ring,* and *little fingers*. In addition, it creates difficulty in flexion and **add**uction of the wrist. The ulnar nerve injury may cause *claw hand* if the elbow fractures the medial epicondyle of the humerus. In this case, the fourth and fifth digits cannot be flexed at the *distal* interphalangeal joints. Also, the individual experiences motor and sensory loss. Remember, the ulnar nerve passes *posterior* to the *medial* epicondyle. In ulnar nerve *entrapment*, the elbow compresses the nerve and causes the medial palm and fourth and fifth digits to become numb.

## RADIAL NERVE INJURY

The radial nerve can be injured following a deep wound to the forearm. This disrupts the thumb and metacarpal-phalangeal joints—they will be unable to *extend,* although sensation will remain intact. The only exception is if the *superficial branch* of the radial nerve is affected. By compressing the radial nerve in the middle of the arm, *wrist drop* occurs. This is also known as *radial nerve entrapment*—it can occur when you fall asleep on your arm—the radial nerve is compressed.

## *BRACHIAL ARTERY*

The brachial artery lies in the *middle* of the arm near the *coracobrachialis muscle*. This is *medial* to the humerus. The brachial artery is a direct continuation of the *axillary* artery as it changes its name at the inferior border of the *teres major muscle*. It gives off a branch, the *profunda brachii artery*, which continues with the radial nerve in the arm. The brachial artery divides into terminal branches throughout. When the brachial artery is occluded, there is a loss of function of the muscles, ischemia of the deep flexors (of the forearm), and problems with digit contraction and flexion.

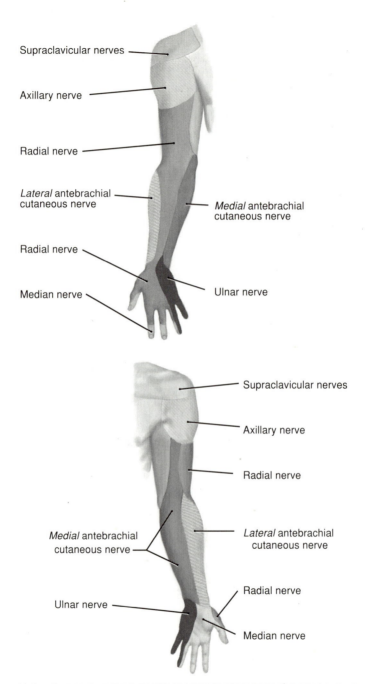

Supraclavicular nerves

Axillary nerve

Radial nerve

*Lateral* antebrachial
cutaneous nerve

*Medial* antebrachial
cutaneous nerve

Radial nerve

Median nerve

Ulnar nerve

Supraclavicular nerves

Axillary nerve

Radial nerve

*Medial* antebrachial
cutaneous nerve

*Lateral* antebrachial
cutaneous nerve

Ulnar nerve

Radial nerve

Median nerve

**Figure 11-8** Cutaneous nerves: posterior (dorsal) view (top) and anterior (palmar) view (bottom).

## RADIAL ARTERY

The radial artery begins in the *cubital fossa* and goes through the anterior forearm to the distal forearm at the end of the *brachioradialis* muscle. It passes through the anatomic *snuff box*, which is bounded by the extensor pollicis longus and brevis tendons. The radial artery reaches the palm of the hand and gives a branch to the thumb, as well as to the *deep* palmar arch.

## ULNAR ARTERY

The ulnar artery also begins in the *cubital fossa* and goes behind the flexor digitorum superficialis muscle to give rise to the *common interosseous artery*. The ulnar artery lies *superficial* to the flexor retinaculum and enters the hand to form the *superficial* palmar arch (it is NOT within the carpal tunnel).

## AXILLA

The axilla is the area commonly known as the underarm. The space is bounded by the *pectoralis major* and *minor* (anteriorly); the *subscapularis*, *teres major*, and *latissimus dorsi* (posteriorly); the *serratus anterior* (medially); and the *intertubercular humeral sulcus* (laterally). The axilla contains several lymph nodes, vessels, and nervous plexi.

## SHOULDER JOINT

The shoulder is composed of the *clavicle* and the *scapula*. At the glenohumeral joint, the scapula is in movement with the humerus. The head of the humerus is stabilized by the *rotator cuff muscles*. These muscles include the *supraspinatus*, *infraspinatus*, *teres* **minor**, and *subscapularis muscles*. A dislocation of the shoulder joint can occur in an injury of the shoulder. The axillary nerve can be injured, causing paralysis of the deltoid muscle and a decrease or loss of the sensation at the shoulder skin.

# Notes

# The Peritoneum, Viscera, and Digestive Tract

# 12

## ESOPHAGUS

Tumors, which are usually found by esophagoscopy, may narrow the lumen and cause difficulty in swallowing. This is known as *dysphagia*. The metastasis from the *abdominal area* is usually to the *left gastric* lymph nodes or through the *thoracic duct* to the venous system. An *acquired hiatal hernia* is fairly common and is associated with increased pressure in the abdominal cavity (versus the thorax) and is apparent with sliding hiatus and paraesophageal hiatal hernia.

*Pyloric stenosis* is hypertrophic (thickened) pyloric smooth muscle and narrowing of the esophageal area to the stomach canal. Thus, the proximal stomach is dilated.

## STOMACH

The stomach forms from the enlargement of the *foregut*. The development rotates the stomach so that the dorsal surface lies *left*—known as the *greater* curvature, and the ventral surface lies *right*—known as the *lesser* curvature. The *left* vagus nerve is therefore created as the *anterior vagus* and the *right* vagus is the *posterior vagus*. The stomach is supplied by the *left* and *right gastric arteries* (along the *lesser curvature*) and the *left* and *right gastroepiploic arteries* (along the *greater curvature*). Carcinoma of the *stomach* is more common among *Scandinavians* and *men*. A partial *gastrectomy* is performed to remove part of the stomach and all the regional lymph nodes (i.e., the *pyloric nodes*). Cancer of the stomach can spread to the *liver*, *pelvis*, and the body through the *thoracic duct* and the *venous system*.

## INTESTINE

### JEJUNUM AND ILEUM

The small intestine portions are composed of the following. The first 40 percent is the jejunum and the last 60 percent is the ileum. The *superior mesenteric artery* supplies branches of intestinal arteries, and these branches form *arcades*, or anastomoses, with each other to form *terminal branches* as the *vasa recti*.

## Ileum Diverticulum

In *Meckel's diverticulum*, finger-like *out*pouches form and create gaps in the intestine. *Meckel's diverticulum* is a remnant of the *vitelline duct* approximately 2 in. in length near the ileocecal junction (approximately 2 feet proximal to the junction). Inflammation of an area of intestine may form *diverticulitis*, which presents similar to appendicitis.

## Volvulus

This is an intestine which has become *twisted* and possibly ischemic or necrotic.

## Colon

The colon is composed of the *cecum, appendix, ascending colon, transverse colon, descending colon, sigmoid colon, rectum*, and finally, the *anal canal*. The colon has longitudinal strings of smooth muscle on the exterior wall; these are called the *taenia coli*. As a result, *haustrae* are formed as enlargements due to the taenia coli.

---

## *SPLEEN*

The spleen develops from *mesodermal* cells in the *dorsal* mesogastrium. Two ligaments, the *gastrosplenic* and the *splenorenal*, join the spleen to the stomach and the parietal peritoneum. Be aware of the location of the spleen as an *upper left quadrant* peritoneal organ—since the *spleen* is a commonly injured abdominal organ (left hypochrondrium location). It may be ruptured through fractured ribs (9, 10, or 11) or other puncture or traumatic injury. It may produce hemorrhage, shock, and can lead to death. Normally, the spleen cannot be palpated since it lies *above* the costal margin; however, if it is enlarged, the spleen may be palpated on the left.

---

## *LIVER*

The liver develops as a *ventral* outgrowth of the gut tube. The liver has four lobes (right, left, quadrate, and caudate). The liver is also an organ that may be injured through a laceration or a fractured rib. It will present with *right upper quadrant* (RUQ) pain. Liver damage is commonly caused by the abused consumption of alcohol. A *liver biopsy* can be performed through the **10**th intercostal space in the midaxillary line. In metastatic carcinoma, look at the *liver* as a common area since the veins are drained here. In a liver transplant, there is a good chance that the liver will take; especially with the use of *cyclosporine* as an antirejection transplant drug. When you hear about a *hobnail liver*, this is a liver with destruction of the hepatocytes (cirrhosis of the liver) and replacement of these cells with firm, fibrous tissue. *Cirrhosis* of

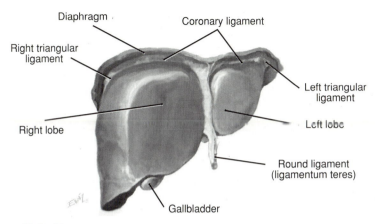

**Figure 12-1**   Liver.

the liver also causes *portal hypertension* with an increased *portal vein pressure*. The liver is shown in Fig. 12-1.

---

## GALLBLADDER

The gallbladder (Fig. 12-2) is located on the visceral surface of the liver and is joined with the liver only by its connective tissue. There are three important ductal systems in the gallbladder: (1) the *cystic* duct (approximately 2 in. in length); from the neck of the gallbladder, the cystic duct joins the (2) *hepatic* duct, at which point it is referred to as the (3) *common bile duct*.

### CHOLECYSTECTOMY

Cholecystectomy is the removal of the gallbladder. Caution has to be taken to avoid severing the *cystic artery*. In the event that bleeding occurs, it is controlled in part by compressing the *hepatic artery*.

### GALLSTONES

An impaction from a gallstone may occur in the constricted area of the ampulla. If the gallbladder expels a stone, the stone may become lodged in the *cystic duct* and create great pain in the *epigastric area*. This is known as *biliary colic*. Another cause of bile duct obstruction is *carcinoma of the pancreas*.

---

## APPENDIX

The appendix may become inflamed and obstructed through *fecal* matter and create an "*acute abdomen*." This includes a presentation of an initial *periumbilical pain*, which then moves to *right lower quadrant* (RLQ) pain. The pain may be referred from the stretching of the inflamed peritoneum with a *rebound tenderness* (if you push in the abdomen and let go, the stretching

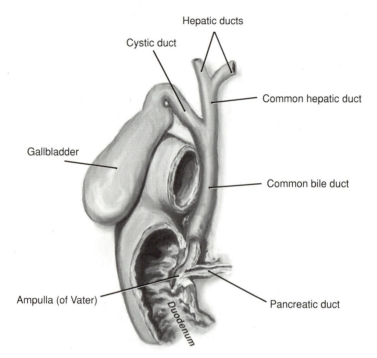

**Figure 12-2**  Gallbladder and ducts.

causes a pain—it's not the greatest way to please a patient). Removal of the appendix (*appendectomy*) is the likely treatment.

## KIDNEYS

The left and right kidneys are located in a *retroperitoneal* position surrounded by fatty tissue and fascia. The kidney is composed of filtration *glomeruli* in the *cortical* region and *collecting tubules* primarily in the *medullary* region. The ends of the tubules form the *renal pyramids*, which flow into the *minor calices* and, finally, into the *major calices*. Finally, these empty into the *renal pelvis* and then the *ureter*. The *suprarenal glands* are also enclosed in the perirenal fascia. They are derived from the mesoderm and are responsible for the secretion of steroids. The *medulla* secretes *epinephrine* and *norepinephrine* through the innervation from the *preganglionic sympathetic fibers* of the *greater splanchnic nerve* (the *cortex* does NOT receive such innervation).

If a patient presents with severe, rhythmic and sharp pain, this may be a result of a kidney stone entering into the *ureter*. This is refered to as *ureteric colic*, and the stones may be formed of calcium, urate (uric acid), and phosphate.

## PANCREAS

The pancreas develops from two gut tube evaginations (one dorsal and one ventral). The dorsal evagination develops into the *dorsal mesentary*. The ventral pancreatic evagination (along with the hepatic bud) develops into the *ventral mesentery*. This ventral bud turns about the gut tube and joins with the *dorsal* bud to fuse and form the pancreas. The inferior part of the *head* of the pancreas and the *uncinate process* form from the *ventral* bud. The *superior* part of the head and the neck, body, and tail of the pancreas all form from the *dorsal* bud. The structures near the pancreas are shown in Fig. 12-3.

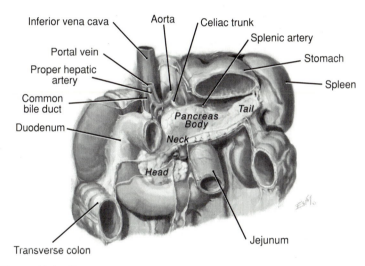

**Figure 12-3** Structures near pancreas.

# Notes

# Index

# Index

The letter _f_ following a page number indicates that a figure is being referenced.

ISBN 0-07-038415-0

90000

9 780070 384156

LINARDAKIS: DIGGING UP
THE BONES: ANATOMY